EVOLUTION OF AN INFORMATION SOCIETY

EVOLUTION OF AN INFORMATION SOCIETY

Edited by
A.E. Cawkell

Aslib

This anthology was produced using optical character recognition techniques and the papers are reprinted largely as they appeared in their original form. References have not been included and the omission of text is indicated by ellipses within square brackets [...].

First published in 1987
by Aslib, The Association for Information Management
Information House, 26–27 Boswell Street, London WC1N 3JZ

British Library Cataloguing in Publication Data

Evolution of an information society.
1. Technology — Social aspects
2. Information storage and retrieval systems — Social aspects
I. Cawkell, A.E.
303.4′83 T14.5

ISBN 0-85142-211-X

Phototypeset using OCR techniques in 10/11 point Plantin
by Inprint of Luton Ltd
Printed and bound in Great Britain by Page Bros (Norwich) Ltd

Contents

Acknowledgements

Information technology and the myth of abundance, Anthony Smith. Reprinted by permission of the author from *Daedalus*, 111 (4), 1982, pp. 1–16.

Knowledge production and occupational structure, Fritz Machlup. Edited version reprinted by permission of the publisher from Chapter 10 in *The Production and Distribution of Knowledge in the United States*, Princeton University Press 1962, pp. 377–383, 386–389, 392–400.

Information and society, Edwin B. Parker. Edited version reprinted by permission of the publisher from Volume 8 of the *Annual Review of Information Science and Technology*, C.A. Cuadra, ed., American Society for Information Science, Washington 1973, pp. 346–356, 365–371.

The post-industrial society: a conceptual schema, Daniel Bell. Edited version reprinted by permission of the author from *The Coming of the Post-Industrial Society*, Basic Books, New York 1973, pp. 112–119, 374–377, 450–455, 487–489.

The knowledge economy, Peter F. Drucker. Reprinted by permission of the author from *The Age of Discontinuity*, Heinemann, London 1969, pp. 247–268.

Artificial intelligence, Joseph Weizenbaum. Reprinted by permission of the author from Chapter 8 in *Computer Power and Human Reason*, W.H. Freeman, San Francisco 1976, pp. 202–227.

What computers mean for man and society, Herbert A. Simon. Reprinted by permission of the author and The American Association for the Advancement of Science from *Science*, 195, 18 March 1977, pp. 1186–1191.

Information: the ultimate frontier, Lewis M. Branscomb. Reprinted by permission of the author and The American Association for the Advancement of Science from *Science*, 203, 12 January 1979, pp. 143–147.

The informatisation of society, Alain Minc. Reprinted by permission of the publisher from Volume 3 of *Policy Implications of Data Networks in the OECD Area*, OECD, Paris 1980, pp. 154–159.

A new concept of information society, Manfred Kochen. Reprinted by permission of the author and the publisher from *The Infrastructure of an Information Society*, B. El-Hadidy and E.E. Horne, eds., Elsevier 1984, pp. 25–38.

The brain organization of the modern world, H.G. Wells. Reprinted by permission of the publisher from Chapter 2 in *World Brain*, Doubleday, New York 1938, pp. 40–57.

As we may think, Vannevar Bush. Reprinted by permission of Dr R.D. Bush from *The Atlantic Monthly*, July 1945, pp. 101–108.

Memex revisited, Vannevar Bush. Reprinted by permission of Dr R.D. Bush from *Science is not Enough*, Apollo Editions, New York 1969, pp. 75–101.

Automata and information, R.A. Fairthorne. Reprinted by permission of the author from Chapter 2 in *Towards Information Retrieval*, Archon Books 1968, pp. 11–21.

Economics of inquiring, communicating, deciding, Jacob Marschak. Reprinted by permission of Thomas Marschak from *American Economic Review*, 58 (2), 1968, pp. 1–18.

Extra-terrestrial relays, Arthur C. Clarke. Reprinted by permission of Electronics & Wireless World from *Wireless World*, 51, October 1945, pp. 305–308.

A mathematical theory of communication, C.E. Shannon. Edited version reprinted by permission of AT&T Technical Journal from *Bell System Technical Journal*, 27, 1948, pp. 379–380, 392–393, 403–405, 407–408, 639–641.

The convergence of computing and telecommunications, David Farber and Paul Baran. Reprinted by permission of David Farber and The American Association for the Advancement of Science from *Science*, 195, 18 March 1977, pp. 1166–1170.

Computers that talk and listen: man–machine communication by voice, James L. Flanagan. Edited version reprinted by permission of the author and IEEE from *Proceedings of the Institute of Electrical and Electronics Engineers*, 64 (4), April 1976, pp. 406–414.

The scientific journal of the future, John Senders. Reprinted by permission of the author from *The American Sociologist*, 11, 1976, pp. 160–164.

Can a machine think?, A.M. Turing. Edited version reprinted by permission of Basil Blackwell from *Mind*, 59, 1950, pp. 433–460.

Social implications of intelligent machines, Margaret A. Boden. Reprinted by permission of the author from *Radio and Electronic Engineer*, 47 (8/9), Aug/Sept 1977, pp. 393–398.

Evolution of an information society

A. E. Cawkell

> Clearly "Information" and its derivatives are words to avoid. It is no more than a linguistic convenience that saves you the trouble of thinking what you are talking about. Fortunately one does not have to use the word. Always, if we put our minds to it, we can say what we mean.
>
> Robert Fairthorne

The subject matter of this anthology of articles and book extracts covers two main topics – social and political change, and developments in information systems and technology. For the most part the articles have been reprinted without editing, or with very minor changes. For books, an attempt has been made to choose the most significant chapter. Mathematics is an integral part of the Shannon article, but with this exception technical parts of articles or books which make reading difficult have been omitted when possible.

In the social change articles, scattered throughout the last fifty years, certain ideas and trends are discussed and there is a good deal of difference of opinion and much uncertainty about the eventual outcome. The ideas were sparked off by analyses of occupations which suggested that the numbers employed in "information" activities are growing at a faster rate than any other. It has been claimed that over half the employed population of the US are "information workers". In reality, many of these people were formerly classified under the less flattering heading of "non-productive" workers — in other words they were an "administrative overhead" required to support "production". Re-labelled as "information workers", the increase in their numbers forms a major part of the growth.

"Information", an intangible entity, is hard to define, and that provides opportunities for deriding the whole notion, just as does the mere re-labelling of occupations. Thus Newman and Newman[1], with information theory in mind, choose the "pure" definition: "Information is that which destroys uncertainty" and quote Braverman[2] who considers that the technical knowledge required to operate the various US Industries is concentrated in only 3 per cent of the working population. The Newmans' definition narrows the field to the extent that those who have been "defined out" of being well-paid "knowledge workers" would, as they put it, be presented with "little scope to identify with the oppressed". I take the view that "information" has broader connotations than in its information theory usage. (It is of some interest to note that ambiguity could have been avoided had "information theory" been called "communication theory" since that is what it is.) Words such as "information", "data", "knowledge", and "wisdom" are often used as if they were interchangeable. "Information" is a very unsatisfactory word and the phrase "the information society" is equally unsatisfactory. In spite of Robert Fairthorne's strictures it will have to suffice to describe a society in which ultimately most people will be engaged in "brain" rather than "physical" work. Few would deny the onward march of this trend.

Whether or not the trend will provide further opportunities for the capitalists to exploit the workers (as implied by the Newmans), there has already been an increase in employment, and in the US a very large increase, in new industries where information or an "information-exploiting" product is the end-product itself. Developments in information technology and communications are affecting both the numbers of people employed and the nature of their work. The work may be highly absorbing or it may be boring. It is sometimes forgotten that the largest occupation created by the advent of computers is that of the data-entry operator. The societal changes which may occur as this trend gathers momentum have given rise to much speculation about whether the information society will turn out to be utopian or divided. The evidence of increased employment by reason of the demand for the new products provided by the "information industry" for "general industry" is accompanied by evidence at least as strong, if not stronger, that these products will be for the use of a smaller labour force with higher skills than were needed by users of the products which they replaced.

The prospect for the displaced lower-skilled people in developed countries does not look very good. For example, "Enlightened management feels obligated to provide workers with a safe workplace ... and supports the human need for comfort, for it knows that productivity suffers when a workplace is disagreeable. Beyond meeting these basic needs, management's responsibility ends. Few organisations are concerned with the long-range impact on the workplace and its technologies on higher-level human needs. This they leave to social, political, and economic forces. If this were not the case management would less readily shift its basis of manufacturing to nations where the costs of labour are low and the prospects for profits high."[3]

A second area of interest, but one which has received rather less attention although more important in the long run, is the possible creation of two societies – the information-rich and the information-poor and the distribution of power associated with this concept. Johnson[4] puts the notion rather well: "One may have power by virtue of natural ability, position, financial circumstances, education, etc. There is no ethical principle that requires that every individual have equality of all types of power. Rather, we aim at equality in certain specific realms. For example, we aim at equality of opportunity in education and jobs, and we aim at equality under the law. Whether or not computers help or hinder the achievements of these aims depends on how they are used." And: "The new technology has not been used to make services cheaper and better for ordinary people. Big business, the better professional firms and such like have been able to contract out of deteriorating services. When I complained of the postal services to a fellow sufferer she replied, 'Of course my husband's firm send anything urgent by motor-bicycle or telex'. This is not an option open to most of us."[5]

There are few comments in the literature about national differences in the onset or nature of the information society, or about local modifications that may occur. It seems to be assumed that the US will lead, to be followed by other countries with a delay and at a growth rate depending on their current state of economic development. There obviously will be great differences for traditional, technical, and other reasons. For example, in France over a million Teletel terminals have been distributed free or at a price far below cost to act as electronic telephone books. Despite some telecom problems the growing customer base has encouraged other services to ride on the investment. This policy decision may result in a widespread familiarity with home terminals in contrast with countries without such a policy.

For the marketeers, of course, there is no uncertainty. Non-adopters of the latest executive four-barelled turboencabulators are simply fossilised. This ebullience is frequently misplaced because of the "conceptualisation" problem. Market research is difficult because for the layman the implementation, usefulness, and pay-off of an information technology application is often extremely difficult to assess. "The marketplace as a determinant of demand in private sector terms depends essentially on the public's familiarity with the product being offered. The demand for a traditional product can be relatively realistically estimated but it is not easy for members of the public to respond in a meaningful way to the typical market survey probe about their interest in new products with which they have no experience."[6]

In the information technology articles included in this volume, some of the confidence of the new technology marketing people has rubbed off on to those who discuss and propound technical advances more thoughtfully. Others are concerned mainly with engineering factors or feasibility and their articles are unaccompanied by sales superlatives. These people are bound to be wrong sometimes about the prospects for particular devices or systems, and they are quite often wrong about the rate at which ideas will become a commercial success and permeate the market, but they are right in assuming a certain inevitability about technical advances.

There is just as wide a range of articles from which to select for inclusion in the second half of this book, as there is for the first. I had no difficulty with the oustandingly inventive; others were chosen because they explain major developments particularly well or raise interesting consequential issues. Of the continuing interest in all facets of the information society there is no doubt. Two quite different aspects have been well set out recently by Miles[7] and Cronin[8].

The work of the selected authors

I hope that the extracts, for which space has been found here, speak for themselves, although it is hard to capture the flavour of a book with a single chapter. The notes which follow group the chapters, explain why each item has been chosen, and summarise or highlight certain parts.

Smith: The media; information abundance for affluent societies

The first article in this anthology is well placed because it eliminates a need sometimes fulfilled by the editor – the provision of a review of historical developments and current trends. Here it fills another gap as well; most of the other articles covering societal changes deal with work and leisure. Anthony Smith is concerned with changes in the media and the way in which information is disseminated. From his work at the BBC and currently as Director of the British Film Institute, his perception of "the dependence of the structure of change upon a number of giant corporations as they grope toward the trend while trying to influence it" demand our attention.

Smith remarks that "in spite of the atmosphere of feverish change that has always beset the information media ... each new development has required about fifteen years to become established within the market". This is a point which seems to need constantly repeating. I have already referred to the information technology marketeers from whom we get the "hurry, hurry, in case you get left behind" breathless message, also constantly repeated. Quoting from some remarks made elsewhere[9] I would remind you that "it usually takes at least ten years from first announcement of a new development to medium scale use. Packet switched networks took 15 years, optical fibres 14, satellites 30. The electronic journal was conceived in 1976 and no viable product is yet available. Continuous speech recognition may not be here for another 20 years".

Smith concludes that "Western societies are on the verge of developing systems ... which will tend to negate the basic principles through which information has travelled through society since the Renaissance ... the whole expansion of the information economy hinges on the consumer economy ... and everything conspires against the society that has a deficit in its national balance of educated talent".

Machlup, Bell, and Parker and Porat: The evidence?

Fritz Machlup (1962) must be credited as the founding father of the "Information Society" idea, although the work of Daniel Bell, and Edwin Parker and Marc Porat had a greater impact. Crude as was the available information and over-broad as was the definition of an "information-worker", Machlup and Porat are almost invariably cited and uncritical assumptions are made based on their statistics in current writings about the information society. These statistics purported to show that up to half of the US labour force was engaged in "information work".

The statistics have been refined in the lesser known work of Debons et al[10], and Schement and Lievrouw[11]. Debons estimated the number of US "professional information workers" to be 1.64 million – about 2 per cent of the labour force. Schement and Lievrouw, defining information workers with the aid of the *US Dictionary of 20,000 Occupational Titles* in a much more recent assessment, came up with a figure of 40 per cent as the percentage of information workers in the US labour force. Again, the actual figures obviously depend on the definition adopted.

Machlup states at the end of his chapter: "What has been said about changes in the occupational structure remains valid even if the basic statistics are somewhat off the mark." He starts and ends the chapter with some remarks about employment, concluding that without drastic reforms in the educational system the prospects for the unskilled are not good.

The chapter I have chosen from Parker's works (1973) is more readable than either his OECD report jointly with Porat,[12] or Porat's massive compilation[13]. Having agreed with Machlup about the key role and growth of information in the economy, Parker discusses the need for efficient investment. He sees the best potential in providing information resources for the economically and culturally deprived. Claiming a "considerable elasticity of demand for information" he suggests that the argument that such people do not want and would not use information even if cheaply available, is flawed; it would be similar to arguing against the widespread diffusion of print following Gutenburg's invention on the grounds that it would not be used by an illiterate public.

Bell, in his chapter (1973), makes a rarely found point about internationalism (referred to in my introductory remarks). "Societies ... may confront a similar core of problems. But this is no way guarantees a common or like response ... which will be relative to the different political and cultural organisations of the specific society." His idealism comes through in the part headed *Just Meritocracy*. He hopes that the information society will include "a meritocracy made up of those who have earned their authority. An unjust meritocracy is one which makes these distinctions invidious and demeans those below ... and there is no reason why the principle of meritocracy should not obtain in business and government as well."

Drucker, Weizenbaum, Simon, Branscomb, Minc, and Kochen: The social implications

These chapters are selected from the works of well known formulators of opinion about a future information society.

Drucker (1969) makes the bold assertion that "the systematic and purposeful acquisition of information rather than 'science' or 'technology' is emerging as the new foundation for work, productivity, and effort throughout the world" and continues: "We learned in World War II with respect to manual crafts that we could compress years of apprenticeship into weeks, or at most months of organised and systematic learning ... average people became skilled craftsmen in little time." Drucker continues with a history of the "rise of the knowledge worker".

The debate about the possible net effect of computers upon society is well covered in the chapters by the next three authors where it is Weizenbaum (the MIT professor) versus Simon (professor at Carnegie

Mellon) and Branscomb (IBM vice-president). Although Weizenbaum is a professor of computer science he has become increasingly cynical about the benefits of a computer-based information society. Many of Weinzenbaum's remarks belong with the last chapters in this book about computers and "thinking". Simon, on the other hand, feels that men will learn to apply computers, with suitable safeguards, to beneficial purposes.

Branscomb makes several technical projections based on the rate of progress up to 1979 but draws particular attention to the kinds of friendly software that must be developed. Like Simon he is an optimist and concludes with expressions of hope for an "information for all" society. Such is the rate of change in technological developments that two areas receiving special mention – Josephson super-conducting junctions and Bubble memories – have not fulfilled Branscomb's expectations. Remarks about voice recognition are also unduly optimistic but there can be no doubt, eight years later, about the continuing rate of advance in many of the areas discussed. In the latter part of the article the possible impact of computers with universal communications free from bandwidth/speed limitations is considered. Branscomb suggests that personalised education at-home may become more effective than in the classroom – as recognised by McLuhan who said, years earlier: "Children resent time spent in school because it interferes with their education." Although he recognises possible suffocation by bureaucracy, Branscomb obviously feels that we are entering a new age of the well-informed citizen.

Alain Minc wants to be sure that French hopes will be brought to fruition by Frenchmen. It is appropriate that his article should follow Branscomb's because in his report with Simon Nora[14], particular attention is called to the need to limit IBM's penetration in France. This report was preceded by an equally influential publication[15] in which Servan-Schreiber suggested that unless the EEC got its act together, Europe would become a subsidiary of the US.

The shorter piece by Minc reprinted here covers similar ground to the report. Minc summarises his views on databank sovereignty, economics, and the societal requirements for France.

Manfred Kochen points out that a copier is knowledge-generating, and a healer carries on a knowledge-based activity which is not knowledge-generating. In his view, an information society is one in which at least half its members are engaged in *knowledge occupations* which are knowledge-intensive as well as knowledge-generating and knowledge-based – for example lawyers. Activities which are not primarily money-earning can also be knowledge occupations. This definition would eliminate many of the Machlup/Porat categories and many years could elapse before the requisite numbers of knowledge occupations are generated. Kochen is also the most idealistic – he suggests that a degree of cultural development is needed during which humanity remains human in the face of dehumanising pressures, and that may take several decades.

Readers may conclude that while something called the "information society" may be on the way, there are a number of opinions about what it will be, or should be, and when it will arrive.

Wells, Bush and Fairthorne: Information storage and retrieval

These chapters are by authors whose ideas about recording and retrieving information have been influential. The steadily increasing volume of published information, particularly scientific information as so graphically outlined by De Solla Price[16], together with the availability of computers programmed to store and select, has prompted many ideas and encouraged the development of numerous systems for the purpose.

However, the proposals made by H.G. Wells for a "World Encyclopaedia", were first put forward in the pre-computer age (1936) during a lecture at the Royal Institution. The chapter reproduced here is the text of a similar lecture he gave during a visit to the US in 1937. Wells found that the "World Knowledge Apparatus" was not up to scratch and suggested that his encyclopaedia, "constituting the material beginning of a real World Brain", should "undergo continual revision ... a perpetual digest and conference on the one hand and a system of publication and distribution on the other". Wells noted that the British Museum Library was microfilming rare books and that "cheap standardised projectors offer no difficulty" – a minor prophesy which has not been fulfilled nearly 40 years later. Sadly, neither has his World Brain – an idea which is still widely remembered.[17] Much earlier (1902), Wells had written a book entitled *Anticipation of the Reaction of Mechanical and Scientific Progress upon Human Life and Thought* containing some correct predictions, but (as at 1986) he wrongly anticipated that increased information flow would bring about improved rationality and justice.

Vannevar Bush, an electrical engineer active in research during World War I, and Director of the US Office of Scientific R & D in World War II was well qualified to suggest his "Memex" machine in 1945 – the first realisation that computers might be the means for a leap forward to something totally new. Bush visualised a translucent desk-top which served both for microfilming images of documents placed on its surface, and for displaying recalled microfilmed images by back-projection. He proposed an "associative indexing" technique for his system so that "when numerous items have been joined together to form a trail, they can be reviewed in turn, rapidly or slowly, by deflecting a lever like that used for turning the pages of a book." "Associative indexing", as opposed to "hierarchical indexing", today usually means the provision for retrieving documents having the same or similar terms in common – a much narrower concept. But Bush anticipated that machines able to learn from experience would be developed and in his second article (1969), he describes improved forms of his associative trails. Something like them might be implemented one day.

Robert Fairthorne was also working in the immediate post-war period, but nearer to the basics than Bush. His work at the Royal Aircraft Establishment, Farnborough, was concerned with computers and information and he arranged to have Hollerith tabulating card-based machines installed at RAE. But he is best known for the application of his mathematical knowledge to classification and coding. His ideas on document "marking and parking", as he called it, were not always incorporated into the early document "mechanisation" systems. They would have been a good deal better if they had. Fairthorne is not, however, a "systems *per se*" man; his main criterion is better service for the user. His paper *The Mathematics of Classification*[18] is still mandatory reading providing you take time to dwell upon it. The article reproduced here is rather easier to read.

Marshak: The value of information

The role and value of information have been much discussed by economists in the context of competitive marketing, advertising, and prices. Even so, G.J. Stigler, 1982 winner of the Nobel Prize for Economic Sciences, has said that "it occupies a slum dwelling in the town of economics." In the context of information and knowledge it has received even less study. Information, information services, and libraries suffer in comparison with more tangible activities when funding is required. A procedure called "cost benefit analysis" has been applied in several fields such as motorway building and flood planning control to try and deal with this problem, and it has also been used for information projects. When there is no easily quantifiable output, as with many human activities, the objective is to express the benefits in monetary terms so that planning and funding may receive attention comparable with other projects competing for a slice of the financial cake.

To try and get to grips with the problem a study was commissioned in 1974 by what is now called the British Library R & D division. The report by Flowerdew and Whitehead[19] is a model of clarity on the subject. Paradoxically its success lay in the manner in which it convincingly showed that evaluating information is extremely difficult. The writers concluded that "no really satisfactory cost benefit study has yet been carried out." The chapter provided here by Jacob Marschak, a well known economist, covers many aspects of this difficult subject. It explores several facets in a most interesting way. The difficulties are evident and Marshak concludes with a number of questions, most of which still remain to be answered.

Clarke, Shannon, Flanagan, and Faber and Baran: Information technology and telecommunications

The articles selected for the information technology section of this book cover several of the most important areas of the subject – telecommunications and computers, speech recognition, electronic publishing, and artificial intelligence.

Arthur C. Clarke's article is remarkable for its predictive accuracy. Noting the performance of the German V2 rockets, Clarke says about space stations: "Although such an undertaking seems fantastic, it requires rockets only twice as fast as those already in the design stage." Clarke realised that the only way to obtain reliable world-wide radio communications was to fire a satellite above "escape velocity" and steer it into orbit. Three satellites could "see" the whole of the earth's surface, and very high frequency radio waves from them would pass easily through the ionosphere with transmitter powers of less than one kilowatt. At the correct speed and distance the satellites would become "geostationary" – they would rotate with the earth, appearing to hover above the same spot on the surface. Clarke has lived to see his "fantastic undertaking" come true.

Claude Shannon's theoretical exposition *A Mathematical Theory of Communication* establishes the criteria for the speed and accuracy with which a telecommunication channel can convey reproduceable messages. Obviously a maximally efficient high speed channel is of economic interest to communication engineers, but why has this article about a seemingly mundane topic had an impact which arguably exceeds any other in this book? In arriving at his conclusions, Shannon was concerned with correctly received messages, not their meaning. He took account of factors affecting the probability of the correct reception of a message including noise, redundancy, and methods of coding. Several parts of his paper have had at least as much, if not a greater impact than the final conclusions about channel capacity. Many of the 78 pages of this article contain mathematical equations so I have chosen to reproduce only the introduction, a part covering probability, and the part concerning channel capacity.

Shannon's work gave rise to information theory – suggested as being the *only* theory in the empirical laws of information science. In fact the work is not about information in the "usual" sense at all. His topic is communication theory – the theory of transmission and reception of symbols via a communication channel. During the 1950s (and even today) this was not properly understood and Shannon's ideas were unjustifiably applied to a whole range of information functions. However, the equations describing the probability of receipt of correct symbols, divorced from "meaning" or "information" in the usual sense, can be usefully applied to the *arrangement* or *frequency* of meaningful data. For example, a word which occurs very rarely, such as "triskaidekaphobia" (abnormal fear of the number 13), has a high "information content", but "high frequency" words such as "and", "the", etc., have a low "information content". The occurrence frequency of words has important consequences in indexing systems.

The conversational computer Hal in Kubrick's film *2001* together with well publicised items like the Kurzweil reading machine foster the idea that voice-controlled computers will be with us very shortly. Enormous interest is excited by such machines and the development of a cheap machine capable of speech dialogue would, of course, be a very important advance. In fact, voice synthesis is easy compared with voice recognition, large-vocabulary word-recognition is more difficult, and limited-vocabulary single-person-generated continuous-speech recognition after a man-machine training session is harder still. But any-speaker continuous-speech unlimited-vocabulary recognition is a technology which could be here in the next century but probably not by 2001.

Flanagan's 1976 review was written during a period of intense research which has continued ever since. At that time he considered that "the speech typewriter ... accepting unrestricted connected text from any speaker may not happen this century or ever." Flanagan heads a team at Bell Labs – a prime centre of excellence. However, with advances in fast parallel processing and with roomfulls of components now compressed onto single chips, perhaps that day has come a little closer. In June 1986 IBM announced that a team led by Jelinek had run an experimental system capable of recognising slow separated-word continuous speech composed from a 5,000 word vocabulary, uttered by a speaker after a 20 minute training session. The significant factor is that it was done using a PC AT microcomputer with a new 30 million-operations-per-second chip, special circuits, and storage of only 640 Kbytes.

Farber and Baran recognise the technical and political dilemmas confronting the policy makers who must struggle with the development of the infrastructure needed by the information society. The sentence "More GNP is developed in the US today in activities in the information sector than in the production of tangible goods" indicates how firmly this idea had been accepted without question even back in 1977. The de-regulating and privatising legislation, since enacted in the US and the UK, is an attempt to resolve a major aspect of the dilemma. It is too early yet to judge its effects. Earlier, Baran was an influential figure on the US Cable/TV scene. He correctly diagnosed problems and assisted with the legislative solutions which enabled Cable to take off in the US.

Senders: Electronic publishing

Electronic publishing means either the electronic distribution of information or the production of print-on-paper with the aid of computers. John Senders is usually credited with the electronic journal idea or more particularly the electronic scientific journal idea. The impetus came from the problems that small publishers, such as learned societies, were having with ever increasing production costs, higher prices, and falling subscriptions. Electronic journals could have been one of the products of co-operative editorial processing centres – an idea which got to the experimental stage in the US but never caught on. The best known "electronic journal" was produced as part of a venture run at the New Jersey Institute of Technology by Turoff and Hiltz[20], although it had little resemblance to a printed product. Further experiments have been carried out in the UK at Loughborough University.[21] The conventional printed scientific journal will be hard to replace as it fulfils current awareness, archival, browsing, and priority functions and is an important component in the sociology of science. The experimental work has been aimed at certain aspects of scientific information exchange particularly by "online conferences".

Since storage costs have dropped a number of publishers have started to provide full text online versions of their printed publications via computer database services such as Lockheed Dialog. The several purposes of this exercise do not seem to include profit-making at present. Reading text from present generation displays compares unfavourably with reading it from printed pages.

Turing, Boden: Intelligent machines

There is a great deal of interest in the achievements of the computer poineers Turing, Von Neumann, Eckert, Mauchly, Wilkes, Williams, and others. Turing is the subject of much curiousity. He led the life of a social deviant and committed suicide in 1954 at the age of 42. In 1936 he read an extraordinary paper to the London Mathmatical Society describing his Universal Machine – a stored program computer – and in 1945 proposed a complete design for such a computer (ACE). His work at Bletchley Park during the war included decoding messages encoded on the German ENIGMA machine.[22] It is of some interest to note that Turing may have been the first person to have had the idea of a stored program machine, and may have discussed it with Von Neumann who is usually credited with being the founding pioneer. Some would say that the Colossus computer, built at Dollis Hill in 1943 and used later at Bletchley, ante-dated all other programmable electronic computers. Because of enduring obsessive British secrecy little was known about Colossus for another thirty years.[23]

The Turing paper reproduced here is probably better known because his 1936 paper was strictly for mathematicians. The question "Can a machine think?" and the more recently coined phrases "artificial intelligence" and "expert systems" are associated with unease about the de-humanising effect of machines and even to the possibility of a "machine-takeover". Turing argues cleverly about the question of machine "consciousness" ending, somewhat unsatisfactorily, with the statement, "I do not think these mysteries necessarily need to be solved before we can answer the question with which we are concerned in this paper."

I find the remarks in the Weizenbaum chapter, mentioned earlier, more satisfactory. Weizenbaum discusses the complexity of different behavioural patterns transmitted from one generation to the next in consequence of being, say, either American or Japanese – "the sort of knowledge acquired with the mother's milk and through the whole process of socialisation that is itself so intimately tied to the individual's acquisition of his mother tongue. It cannot be learned from books; it cannot be explicated in any form but life itself ... what could be more obvious than the fact that, whatever intelligence a computer can muster, however it may be acquired, it must always be absolutely alien to any and all authentic human concerns?"

Margaret Boden was the first person to take a cool look at "thinking machines"[24] marking the end of a period of heady optimism and speculation which followed McCarthy's coining of the term "artificial intelligence" in 1957. The article which is reprinted here was written shortly before her book was published. Having reviewed the then current achievements in "intelligent machines", she discusses the many comments which have been made about "dehumanising effect" as "people become decreasingly dependent upon human contact for satisfaction and their needs", and continues with some remarks about welcome effects – for instance in education. Her remarks are on quite a different plane from earlier pie-in-the-sky ideas such as pupil-teacher interaction by brain-waves – the speculation of the period was epitomised in reference 25. Boden also gives some examples of harmful effects which could be built into programs and proposes that "programs should be written in a sensible fashion so that the control structure is perspicuous and alteration is facilitated." In her opinion the computer profession will need to recognise the issues and take appropriate measures to reassure the public. I have to say that eight years after this article was written I cannot see much evidence that the profession is adopting this attitude.

I will conclude on this note of possible "dehumanisation" with some remarks by Van Cott[3] which tellingly bring the issue to life. He quotes recent comments by Harvard psychologist B.F. Skinner who "... assails the trend towards assigning institutions and machines the work individuals once did. Instead of confronting a noisy neighbour directly, for example, people now ask the police to intervene. Even the task of washing dishes by hand is an example of the small tasks that once filled people's lives with a sense of achievement, co-operation, and camaraderie. These direct experiences of 'doing and seeing' – positive results from human interactions – are greatly to be valued. Their loss in the home and workplace is causing a sense of alienation and boredom ... while the human body may not be violated in a workplace of passive monitoring and supervision, the human spirit is."

References

1. Rhona Newman and Julian Newman. Information work: the new divorce? *British Journal of Sociology*, 36(4), 1985, 497–515.
2. H. Braverman. Labour and monopoly capital. *Monthly Review Press*, New York, 1974.
3. Harold P. Van Cott. High technology and human needs. *Ergonomics*, 28(8), 1985, 1135–1142.
4. Deborah G. Johnson. Mapping ordinary morals on to the computer society: a philosophical perspective. *Journal of Social Issues*, 40(3), 1984, 63–76.
5. Jo Grimond. Why is technology not helping the common man? *The Daily Telegraph* (London), 6 January 1986.
6. Mary G. Jones. Citizen's rights to information: the role of government. *Information Services & Use*, 5, 1985, 37–47.
7. Ian Miles. The new post-industrial state. *Futures*, December 1985, 588–617.
8. Blaise Cronin. The Information Society. *Aslib Proceedings*, 38(4), April 1986, 121–129.

9. A.E. Cawkell, The real information society: present situation and some forecasts. *Journal of Information Science,* 12(3), 1986, 87–95.
10. Anthony Debons, Donald W. King, Una Mansfield and Donald L. Shirey. *The information professional: survey of an emerging field.* Marcel Dekker: New York, 1981.
11. Jorge R. Schement and Leah Lievrouw. A behavioural measure of information work. *Telecommunications Policy,* 8(4), December 1984.
12. Edwin B. Parker and Marc Porat. 2. Background Report. In: *OECD Informatics Studies,* Paris, 11 February 1975, 87–129.
13. M. U. Porat. *The information economy* (7 volumes). Government Printing Office: Washington, DC, 1977.
14. Simon Nora and Alain Minc. L'informatisation de la société. *Report. La Documentation Francaise,* Paris, 1978.
15. Jean-Jacques Servan-Schreiber *Le défi americain.* Also available as *The American Challenge* by J.-J. Servan-Schreiber, translated by Ronald Steel. Penguin Books: Harmondsworth, England, 1969.
16. Derek de Solla Price. *Little science big science.* Columbia University Press: New York, 1963.
17. Eugene Garfield. The World Brain as seen by an information entrepreneur. In: Kochen M. (Ed). *Information for Action.* Academic Press: New York, 1975, 155–160.
18. Robert Fairthorne. The mathematics of classification. *Proceedings of the British Society for International Bibliography,* 9(4), 1947, 35.
19. A.D.J. Flowerdew and C.M.E. Whitehead. Cost effectiveness and cost benefit analysis in information science. *Osti Report 5206,* 1974.
20. Starr Roxanne Hilz. Recent developments in teleconferencing and related technology. In: Cawkell A.E. (Ed). *Handbook of Information Technology and Office Systems.* North Holland: Amsterdam, 1986, Chapter 46.
21. B. Shackel, D.J. Pullinger *et al.* The BLEND-LINC project on electronic journals after two years. *Aslib Proceedings,* 35(2), 1982, 77–91.
22. Andrew Hodges. *Alan Turing: the enigma.* Hutchinson and Burnett: London, 1983.
23. Anon. Was Colossus the first computer? How secrecy lost Britain the lead. *Computer Weekly,* 28 July, 1977.
24. Margaret Boden's article, in revised form, is to appear in the second edition of her book '*Artificial intelligence and natural man*' (first edition 1977) to be published by MIT in 1986/87.
25. G.B. Leonard. *Education and ecstasy.* Delta Press: New York, 1968.

Information technology and the myth of abundance

Anthony Smith

> Inevitably, the culture within which we live shapes and limits our imaginations, and by permitting us to do and think and feel in certain ways makes it increasingly unlikely or impossible that we should do or think or feel in ways that are contradictory or tangential to it.
>
> – Margaret Mead
> *Male and female*

What really *is* a technological revolution? And who are the revolutionaries? What are the criteria for a historical process to be so described? Where do we *look* for the results? I cannot claim that this essay will answer these questions, nor even that they are precisely enough formulated for useful answers to be provided. But it might be possible to construct a new kind of mental picture of the phenomenon of technological transformation, of its driving motivations and cultural consequences, by looking at certain aspects of current developments in information technology within a historical setting. The advent of printing had obviously "a great deal to do" with the rise of the nation-state, and Elizabeth Eisenstein's researches and narrative, as well as those of Martin and Febre, have helped to fill out the picture of a late Renaissance "information revolution." An age in which a new transforming technology is taking hold must, almost self-evidently, express its most profound social, economic, and political changes in terms of that technology – so closely and complicatedly, that historians inevitably try, but fail, to disentangle the resulting skeins of cause and effect. Was there a drive for empire that altered the technologies of European navigation in the fifteenth and sixteenth centuries? Was imperialism a result of technology-push or economic-pull? Do the processes of "take-up" of innovation relate to the dominant creative, emotional, and intellectual mindset of an age, so that the tracks may be found again through later research?

Such questions are the permanent concern of social history, but need to be asked also in the present, while the transforming process is underway. The intention of this essay is to suggest that there exists a great unifying social and cultural urge behind a technological revolution, particularly one that relates to information. The search for the emotional satisfactions of the vernacular and the evolution of the feeling of nationhood were indeed tied to the technology of "moving letters." The Victorian bourgeois' obsession with the perfect mechanical reproduction of images in movement, sound, and hue had some psychic link with the evolution of the representational technologies of film, phonograph, telegraph, and the wireless. Today, a surging belief in the perfect development of the individual as consumer is somehow discovering its own confirmation in the development of technologies of information abundance; but that individualism is tied also to a new imperialism or transnationalism in the growth of the phenomenon of cultural dependence of the South upon the North. The new information technology is reconfirming the world vision of the developed world, reestablishing its confidence as the primary subject of culture, as the developing nations fall victim to the cultural pressures of external data flow. Thus, if this argument works, information technology leads toward the displacement of nationhood, of national cultures.

We have all become modishly aware that the information environment, so to speak, of the late twentieth century individual is in the course of being transformed. News columns with titles that play neatly on the words *revolution, age, galaxy, shock,* appear monthly. But we remain prisoners still of an essentially Victorian idea of the requisite constituents of social change, in the sense that we tend to predicate the transformation upon the technology. We relate and chart development according to a measure of machinery, alongside the evolution of inventions. So numerous are the gadgets of the computer age (there goes an example of what is being criticized!) that the designated historic turning points – the number of "revolutions" per decade – are too numerous to absorb, their effects too shrilly predicted for easy listening. We are paralyzed by the dimensions of the transformation, partly because we have internalized a kind of Whiggian principle, by which machines "produce" social effects of a measurable or, at least, observable variety. The trouble is that technological and social history cannot be related in this way, since the extrapolated trends shoot off the graph every time. Consider the influence of the photocopier, the coaxial cable, television news, teleconferencing, and so on. There are no anchors to cast in each voyage of speculation; every trip rushes straight toward infinity.

We would be greatly helped in the present epoch of speculation if we had available some improved metaphors for social change, something less traumatic and less overworked than "revolution," something more intermingling of cause and effect, something that suggested less emphasis on technology and placed more pressure on social need as the starting point of technology. New technologies close gaps, resolve tensions, register the temporary shelving of problems, as well as automate jobs out of existence and fill the home with new junk. Above all, in order to reduce the current bewildering hyping of technical history, we need some explanatory models of the inventing process that demonstrate the collective, though concealed, social dialogue that almost invariably precedes the advent of a new device.

The apparatus of the modern media of information has been accumulating steadily through the century; the modern home may possess a telephone and a typewriter, a camera, a record player, a pocket calculator, a pile of disks or cassettes, probably by now a couple of television receivers, possibly a cable TV link, an 8mm film camera and projector, a device for playing video games, for receiving pay TV, for decoding teletext or videotext signals, perhaps even a video camera, and a home computer. Few people, however, are as yet aware of the linkages that exist – or that can exist – between all of these gadgets; the information revolution of the late twentieth century consists very largely of the increased propensity for these text and moving image machines to converge and to interact. That propensity has been latent since Victorian times. Thomas Edison invented the phonograph as a repeating device to aid the telephone, thinking that a central office such as the telegraph bureau would record messages sent down the telephone lines and deliver the disks to the homes and businesses of nonsubscribers. It was not a fallacy so much as a prophecy, for we are witnessing today that intermixture of telecommunications with information storage that he envisaged. What happened in the late 1970s was a sudden increase in the potency of telecommunications and in the computing capacity of society that has made it possible for us to reap a whole series of benefits that were impossible when the same technical possibilities existed on a smaller scale. The present "revolution," if such it is, is one of investment rather than technical innovation, of transformation of scale more than of technological horizon.

All of the devices that have emerged as discrete physical media of information and entertainment have their own industrial housing, so to speak. The century has witnessed the growth of a music industry around the phonograph and radio, a TV industry, a film industry, and telecommunications, computing, and book publishing industries. These great blocks of investment and industrial activity are currently undergoing a transformation, and in every society in which they flourish (surprisingly few, in fact, since most societies are becoming highly import-dependent in respect of media software), there is currently a reconsideration of the regulatory environment in which they operate. In some societies, the process is being labeled "deregulation," where it is perceived as a process of removing legal constraints against intercorporate competition. In other societies, the same process is envisaged rather as one of making new and appropriate regulations to stimulate similar releases of enterprise, often accompanied by moves to protect indigenous culture. The new devices of cassettes and disks and the new paid broadcast services entail an extremely complicated regearing of all the established industries that hitherto have been device-specific. In other words, it has to be possible for a set of rights and obligations that have been acquired in respect of a given artifact (say, a film made for theatrical release) to be transferred to a wider range of distribution systems (cassettes, cable television). The changing situation is bringing about a gradual alteration in the way we think about the property element in information and entertainment, and about the cultural demarcations between genres. At one level, the change consists in a series of publishing devices and promotional arrangements; at deeper levels, it must alter our ideas about what constitutes a "book," what separates an "academic work" from a popular one, indeed what body of data should properly be considered a book or an "author." Let us consider a not unusual career for a modern work of fiction. It may begin as a novel about which an individual writer has pondered for years, or it may originate as a commission conceived by an agent or a publisher and fostered upon a writer of recognized skill. If it seems likely to sustain the investment, the finished work may be promoted, and through dextrous manipulation of the apparatus of literary review and public discussion, forced through a series of different kinds of text distribution. It will come out in hardcover and paperback, in serial fiction and digest form, and then as an even cheaper paperback. But it may also be transmuted into a set of moving images, where its basic authorship will be further dehydrated and industrialized in complex ways. A film designated for cinema distribution may in fact be shown, in widescreen format, only for further promotional purposes; the 70mm image will be seen only by a small fraction of the emerging audience, as the work passes into 35mm and 16mm gauges for distribution in various specialist systems (such as the film society network or the college circuit). It will appear in cassette form (all the framing of the original lost in the transformation to the smaller screen) and videodisk, on cable and pay TV, ending up on "free" over-the-air television, public or commercial. At later stages in its career, the work may return to one or more of its earlier phases, but it will remain in public consciousness with greater permanency than that bulk of Victorian fiction which failed to become one of the tiny band of classics.

The new work of today faces a wider variety of audiences and enjoys a more finely calculated career. It is commensurately more heavily dependent upon promotion, and, indeed, more and more different kinds of entrepreneurs will speculate upon its possibilities during the course of its complicated life. There is a rush of newcomers to the marketplace, but inevitably, a wave of cartelization will ensue as soon as this market is rationalized. Thus in this period of convergence of devices, a new division of cultural labor is growing up among them. At the same time, there is a search for new and appropriate forms of material, not dissimilar to that which took place when the telephone and cinematography were evolving, when perception of social role preceded each of a multitude of technical offshoots. The social impact of television and telecommunications has been much subtler and more far-reaching than that of other devices of the same era. Both telephone and cinematography were very slow, however, in gathering around them the aura of transformation that today envelopes the offshoots of the computer and the television receiver.

In the 1880s it seemed possible that the telephone would become a medium of entertainment. In London and Paris, experimenters were to set up connections between the principal theaters and the central operator, so that subscribers could listen to plays and to the songs of the music hall. Others thought that the new medium would be a useful supplier of general information, supplementing that of the newspaper. As a person-to-person instrument, it suffered from obvious limitations: there were few people with whom one could speak, the costs were high, and established systems of social discourse inhibited subscribers from incorporating the machine into their lives. The telephone was neither intimate nor reliably private. It was often confined to small professional groups, such as the lawyers and doctors of Glasgow, who enjoyed their own separate and mutually incompatible exchanges. As an instrument of business, the telephone suffered from other limitations: it was more expensive than using messenger boys; it created tension within the national telegraph administrations (one of which actually proposed charging for the telephone according to the number of words spoken along the line). It came into use at first through a series of specialist groups: the construction teams on early skyscrapers, the police, doctors, lawyers, and so on. It grew within the interstices of society, later coming to occupy a more general public role. The influence of the telephone on the development of social structures and the physical layout of societies is extremely hard to calculate and has tended to be overlooked.

Road and rail systems are more visible, and the great feats of engineering that made them possible have seized the attention of social historians more tenaciously than the invisible forces of telegraph, telephone, and radio. But the areas of influence of the communication devices are themselves different: while transportation facilitated suburbanization in the present century, communications has had a great deal to do with the changing "images" of the different parts of a city, the constantly shifting areas of respectability and trend. Suburbanites have not been migrants on the whole; they have desired to retain the advantages of metropolitan life and to remain in constant touch with the centers of the society, while shunning the geographical core. Communication systems have thus helped with the light-and-shade of social evolution, and have provided a wider range of matters over which social nuance can be expressed. However, each device developed in the Victorian era (and later) began life in an aura of a certain vagueness as to its destined purpose. Was the telephone destined for entertainment purposes, or did it fulfill a special role in preserving social order or in providing general information? Did it facilitate person-to-person or group-to-group communication? The purposes have constantly shifted, although each device eventually acquired its own clear purposes, its own "culture." Now, all is in doubt again, all the boundaries are moving.

Each fresh wave of new devices has registered and expressed a new stage in the evolution of city structure, in neighborhood development, and in the structure of the family. Cinemas in the twenties and thirties released people from their homes; television in the fifties recemented the home as focus of the family, until in the sixties, a new politics of the family seemed to break up that tight postwar grouping. Today, the new multiple devices slice up the family and reindividualize it, permitting and encouraging a new microconsumerism, the pursuit of a fresh, but (temporarily) satisfying, illusion of individual gratification through endless freedom and "choice."

The communication and representation devices of the late Victorians were manifestations of a vast, unarticulated urge; they were an act of ideology expressed again and again, in different versions, of a machine for the perfect reproduction of the lifelike. It is worth dissecting the "invention" of film in some detail and comparing its tortuous progress, via a complex of interactions between technique and social aspiration, with the phenomenon of today, when it is still difficult to express, in one similarly neat phrase, the nature of the parallel contemporary aspiration. There appears today to exist a latent collectivist, egalitarian consumerist urge, a prompting to break through economic and institutional contraints, toward an abundance of messages, from which a mass of individuals can draw material according to their "personal" choice. Choice is the chimera of age, the hypothesis of a new adulthood arising from the opportunity to "perfect" the self as the basic mechanism of consumption. The Victorians, however, were pursuing, through their technologies of illusionism, an ideal mode in which their desire for a kind of artificial immortality could be assuaged.

The moving image was a substitute for – an extension of – the cemetery. The mass suppliers of early film cameras blatantly exploited this deep need of the age: the retention of the perfect images, in motion, of one's loved ones. One of the newspapers that reported the first public screening by Louis Lumière of his films said: "When these cameras become available to the public, when all are able to photograph their dear ones, no longer merely in immobile form but in movement, in action, with their familiar gestures, with speech on their lips, death will no longer be final." The representation and extension through time of the human body was one aspect of the great bourgeois aspiration, but the techniques have their roots in the Renaissance, with the development of anatomical study and the growing importance of perspective as the enabling science that made it possible for the new knowledge to be recorded and imparted.

Marey's discovery of the persistence of vision in the nineteenth century played a similar role in the development of machines to capture, record, and dissect the nature of movement. But Marey was not in pursuit of moving pictures as illusions; indeed, he rejected the machines that enabled movement to be synthesized in favor of another line of invention (which included his own chronophotograph – the first camera to employ celluloid strip) that captured movement on a still frame through strobic effects. Marey and his follower Londe were both attempting to overcome deficiencies in human perception rather than create an illusion of reproduced movement that deceived human perception. They wanted to depict reality in a form that slowed down or speeded up true movement. In a sense, Edward Muybridge's notions lay along the same line of thought; his Zoepraxiscope succeeded in reanimating a series of still photographs taken in succession by different cameras of a horse in motion. The purpose was scientific – to dissect a natural phenomenon and to reconstitute the movement, in order to prove the correctness of the analysis. Muybridge's images were generally taken to be aesthetically unpleasing. They were to influence painting in due course, but only after delivering a shock to those artists of the traditional schools who thought that the real world corresponded to the idealized images that had been taught academically.

Earlier in the century, a flow of devices demonstrated the other half of the Victorian inspiration – illusionism. Daguerre's invention was a natural development of his skill in trompe l'oeil. At the Great Exhibition of 1851, the stereoscope had drawn fascinated crowds, since it seemed to add a dimension to still photography by supplementing a gap in the pure representation of nature. Baudelaire mocked the hungry eyes "bending over the peepholes of the stereoscope, as though they were the attic windows of the infinite." At the same time, the demand for the perfect illusion was being fed through nonphotographic devices for creating the sensation of movement or of three-dimensional images, such as the Thaumatrope and the Praxinoscope, which depended on a disk with images that appeared to combine in motion through rapid rotation. These were representational toys dismissed by the more scientific school of Marey, since they demonstrated, but did not analyze, scientific phenomena. Thus the pursuit of the Victorian aspiration for representing an image of life in its perfection veered from one line of development, in which its various manifestations were deemed to be recreational and at best educational, to another line, in which the quest was not to substitute for painting but to serve science.

It was the Lumière moving-picture show that captured the contemporary imagination in the mid-1890s. Writers as far spread as Gorky, Kipling, and Henry James witnessed the show as it traveled between Spain and Moscow, Austria and America. They were presented with one-minute scenes of real-life events, unstaged but nonetheless contrived, through choice of camera angle, location, and timing. The train arriving at a station, the workers leaving the factory, the gardener turning the hose upon himself – all became familiar metaphors that impressed upon those pioneer audiences the first collectively experienced moving images that inaugurated the great store of shared allusion which has subsequently accumulated. At the same moment, Edison's laboratory and its breakaway group, the American Mutoscope and Biograph Company, were feeding the same popular drive toward perfect representation with a series of inventions that offered short travel or anecdotal films projected in a machine into which the viewer had to peer. Edison was working to a much grander design that led him to overlook the importance of his own silent camera and projecting device; he wanted to link his own phonograph invention with a film and recording technique, with stereoscopic effects added for good measure. Edison wanted to perfect a total system of representation that would perceive, record, and transmit across space, and he thus employed, in a spectacular series of electrical and mechanical inventions, the whole range of sciences that were to be worked on in the course of the following eighty years. All of the experimenters alluded to, and many others, were technicians primarily, but with a keen eye either on the audience of science or on the audience of contemporary showmanship, sometimes a little of both. The first decade of the twentieth century saw a further group of experimenters who were primarily artists, such as Georges Méliès and Edwin Porter, though both had some technical background. Both saw the possibilities of developing narrative forms by means of the new moving cameras. Méliès had started as a conjuror, and he drew from a line of nonrealist illusionism for his film ideas, of which he developed hundreds. His stock – transformations of human beings into animals, the appearance of strange apparitions, a Jules Verne-type space journey – was the culmination of the aspirations of a Victorian conjuror. Porter, however, drew upon a realist line of narrative, breaking away from the theatre rather than from popular showmanship. He is famous for having developed the pacing techniques that laid the foundations of modern narrative cinema – the cutting, linking, and transposition of shots, and the suggestion of simultaneity through building-up a chase scene by intercutting events taking place in different locations.

Behind these different lines of artistic and technical development, all of which proved ultimately interactive and intensely creative of technology, was a growing institution, cinema. Behind each movement forward – and many that proved to be cul-de-sacs – there lay a shared phenomenon that grew by accretion, based upon an audience whose perception and expectations were being progressively intermingled; each mechanical device depended upon a range of artistic conventions that had to be accepted and internalized by a rapidly growing audience if the institution of cinema was to develop. That institution itself was obliged to follow the contours of contemporary taste, to search for the city locations, the distribution systems, the pricing mechanisms, the patent and copyright devices that would sustain the new medium. The technology that emerged registers as a series of interim readings of the relationships that between them constituted, as they still do, the institution of cinema.

Of course, it is impossible to produce a perfect record of the evolution of a technology by concentrating on the interactions and dependencies – social, artistic, technical, and intellectual – since to do so would entail an analysis of an entire society. Marconi's work on radio was taking place at exactly the same moment as the work of Lumière; Zworykin was working on the basic principles of television in the same decade as the main work of Méliès. Regulatory systems for the telegraph and the telephone were being simultaneously created, and these were greatly to influence the early and continuing institutions of radio and television. And of course, contemporary developments in all of the other sciences – from biochemistry to metallurgy, optics to engineering – played their part in the evolution of the communicating technologies. Nonetheless, those lines of development that led to film, television, and radio entailed clusterings of technique that derived their impetus from the same animating aspiration, that of creating a perfect and enduring representation of the perceptual world.

It is rather harder to discern the central drive that unites the various new communication media that have been developed in the last and current decades from that earlier aspiration. Certainly, there no longer exists an unsatisfied craving for the mere illusionist representation of reality; perhaps a reverse principle might today be at work, whereby the techniques for suggesting reality are being pushed toward a realm of perfect fantasy in the new potential for, say, computer imaging. But there is some more general demand or perceived demand that is being stimulated and satisfied by the new media – demand for an abundance of supply and an image of the consumer as individual, arising above an ocean of materials. There is the image of a new leisure, a worklife without toil, a textured, variegated career structure. Many of the new devices are concerned with text storage and distribution as much as with still or moving images, with data processing as much as with storage. Some of the new devices appear, in the present stage of their development, to be concerned with supplying a new multiplicity of channels (cable systems, videodisks, satellite broadcasting, cassettes), while others have more to do with adding to the conveniences of the home or reducing information overload (home box office for first-run movies, videotext), or both. The suppliers of new services are breaking down, unconsciously for the most part, old traditional genres, such as the newspaper or magazine, by offering the chance to dial directly a specialist line of information; they are also providing a chance to evade the many constraints of over-the-air broadcasting with its "paternalistic" overtones of prescribed, preselected patterns of material. The new specialist cable channels are on the whole reworkings of the public broadcasting service model, and offer the chance for a new kind of self-definition on the part of the subscriber into a class or subculture type, rather as the newspaper industry did in the era when newspapers of every conceivable stripe flourished.

All of the new services have broken into preexisting monopolies of some kind, but all are searching for new monopolies of their own in order to survive – monopolies of first-run movies, monopolies of travel or business information, monopolies of high culture material, monopolies over certain geographical zones or certain social groups. With the advent, a decade or so from now, of direct broadcasting by satellite, a wholly new complication will arise, since the satellite, unlike any other transmission system devised hitherto, is capable of equal address across a whole society or group of societies.

Yet these services are, in the main, systems of supply quite separate from the industry providing material to a variety of systems reaching different layers of the audience. Thus a cable offering a cultural channel acquires its content from a multinational industry, programs that have been created to serve a primary market elsewhere (though with an eye to further sales). The market for software is becoming many-layered, even though various homogenizing market forces have already set in. The previous forms of distribution for much of this visual material continue to exist and tend still to be the primary sources of funding – the major national television networks and their independent suppliers. There are half a dozen annual markets and festivals at which the main lines of dealing and the main relationships are built up: Milan, Monte Carlo, Cannes, Berlin, New York, Los Angeles. Film festivals have created video offshoots, with new video markets in Europe and the East being established. But the dealing on individual projects continues throughout the year by means of bilateral arrangements between banks and production houses, television channels and cable organizations, Hollywood majors and publishing houses. Inevitable rearrangements of capital within the media conglomerates are taking place. A new world industry of moving-image products is emerging, highly diverse, but still dominated by the companies that established their grip over the heartland of the audience in the days of the old national television monopoly. The material today is beginning to pass through a complex mesh of distribution systems, each one technology-specific, each with a different pricing mechanism and in a different stage of development. A European publishing house owned by an American bank, for example, will initiate a project designed as a series of films and an international book. It will presell the films to a London-based television company for a price that covers a large proportion of the basic production costs. To cover the rest, it will presell the same series to a U.S. public television station or one of the New York cable stations, the London TV company retaining a percentage. The publishing house will then proceed to organize translations that will sell well, on the reputation of the British and American television transmission. The profit to the original publishing house, however, will tend to come from vastly enhanced book sales, since the whole scope of the market for the books has been transformed by the broadcasting operation. Gradually, the materials will flow into other cable and box office systems, while selling in cassette form in the education market. The product will retain strong national overtones; it is owned ultimately by the American bank, but its makers are British, and its accent will tend to be also. The same thing is happening in Paris, Frankfurt, Tokyo, and Amsterdam, but five geographical locations in the developed world are coming to play an ever greater role in the world supply of moving images and, indeed, of text materials also.

The five locations are New York, California, London, Frankfurt/Munich, and Tokyo, places where there exist strong and sophisticated national audiences for the first-generation television materials, plus the necessary access to capital and the habit of working together on the part of a critical mass of relevant skills and institutions. It is likely that these five centers will remain at the heart of the world market for software in the entertainment and information fields. (Computer translation might eventually enable the Japanese to break also into the world text-information software market, already dominated by Japanese hardware.) There was a similar concentration in the world of book publishing a hundred years after the development of moving type, but the evolution of the newspaper, with its polycentered culture, was quite different. Quite different, too, was cinema, which, though it rapidly became a narrow market, began with a wide variety of supply centers. The developing world contains a number of major centers for film-making (India, the

Philippines, Hong Kong), but these have remained largely national in the 1970s. Today, one or more of them could break into the wider world market, but probably only through major investment from the existing centers, since they are "hampered" by the different musical and literary traditions of the East.

It is clear that two quite distinct developments are taking place. There is a new range of physical artifacts on which are inscribed images and text – cassettes and disks – and these are distributed in roughly the same manner as books and gramophone records. These are, however, to some extent different from their forerunners, in that the material they offer is already familiar to the potential buyer, through the promotion and marketing of a film, of which the cassette or disk constitutes an extended line of supply. The other new media are all services rather than artifacts, although the recipient may, legally or illegally, make a physical copy of the text or image in the home. Thus the new videotext systems are publishing devices, where payment is made through the telephone company for "pages" of material that have been received on a domestic television receiver. Some of the cable systems are paid for overall by the subscriber, as European public broadcasting systems have always been, while other cable systems or scrambled signal systems oblige the viewer to pay for each selected program; these leave the recipient without a physical artifact, unless a domestic personal recording has been made. Policing the uses made by individuals of private recordings is, for the rights-holders, something of a nightmare, and pricing mechanisms are having to adjust for the practical impossibility of retaining rights long-term after the distribution of a new product. The owner of those rights has to consider the timing of the whole package of new media outlets, relying on industry-wide organization for the policing of the multitude of new networks that are springing up.

It is still far too soon to see which technologies will prevail for specific purposes, to discern whether an optic fibre network set up nationwide would eventually take over and swamp all other systems of cables, microwaves, direct satellites, and broadcasting channels, in the establishment of a universal broadband domestic system, a kind of general information ring-main, like the electricity ring-main, linking every individual to the entire national and international system. It does appear, however, that Western societies are on the verge of the development of a system, or collection of systems, that, in their net effect, will tend to negate the basic principles by which information has traveled through society since the Renaissance. Even though the cassette and the videodisk operate in the same mode as books, distributed on the basis of single copy purchase by each user or group of users, the pressure of the nonartifact services is such as to suggest that the artifacts may play a diminished role in the longer term. The Gutenbergian principle is so firmly rooted in our culture that it is hard to imagine a society in which it has been abolished (and, indeed, no one is suggesting that anything like abolition is likely to occur). Rather, we are liable to witness a rapid erosion of the settled notion that information is naturally multiplied in physical copies until the number of copies approximates the number of those wishing to receive it. The Gutenbergian principle has already ceased to function in the case of broadcast material, where the opposite – or what one might call the Alexandrian principle – operates. There, a single copy exists in the originating tape or live performance, which then reaches its audience in nonmaterial form; a physical tape can be generated by the individual recipient, but the mass multiplication of physical materials, as in the newspaper and in publishing, is absent.

One uses the image of Alexandria, because it suggests a great store of material that is deemed to be fully authentic, but available only to those who come to it to choose. A modern data base is, in a sense, an electronic version of the principle where material is added to a central store according to fixed and accepted methods, and is then available to all who have the means and skills to unlock it. In the field of moving images, the world is today steadily building such a store of accredited materials, which have, most of them, been through the authenticating procedures of network transmission before becoming available through the newer systems of distribution. Unlike the materials of a great library or a computerized data base, these materials have still to be laid out as a program by a cable company or satellite distribution company before they can be chosen; but as broadband systems develop, we are veering slowly toward some new condition in which an individual can choose electronic dissemination of a single item that was itself chosen from a vast or total store of video products. In the field of data, this condition is rather closer, if anything, as the various videotext systems slowly agree on international technical standards and interconnections. One further aspect of the steadily dissolving Gutenbergian principle is the part that distance has always played in fixing the cost of communication of any kind: this has applied equally in the case of the telephone and telegraph and the printed book. As the electronic systems emerge, it is becoming increasingly clear that distance is a rapidly diminishing factor in costs, both of collecting information and of redistributing it.

We are witnessing, therefore, a subtle transformation of the underlying principle that has sustained the information systems of human society since the Renaissance. The shift is coming about as a result of a vast number of quite separate responses of corporations to perceived demand, responses of technology to science and of science to imagination. There is no central machine generating this change, no great corporation or conspiracy of corporations. There does indeed exist a powerful, almost total dependence of the whole structure of change upon a number of giant corporations, but they are tending to grope toward the trend while trying to influence it. Their corporate needs greatly influence the pace of change, and while they often choose specific private directions for a period of time, the central pulse reestablishes its rhythm.

Despite the atmosphere of feverish change that has always beset the information media, the basic technologies and content forms have changed very slowly indeed. One may take, for example, the novel and the newspaper as direct emanations of the printing press, and note how each has changed fundamentally in form not more than, say, once in a century. Despite the enormous number of attempted means for creating moving pictures in Victorian times, the celluloid strip, which established itself in about 1897, has remained on the same gauge until today. The development of celluloid only took place at the end of the 1880s, and the earliest cameras for shooting a succession of images on a moving strip of celluloid hardly left their experimental stage before 1895. And yet a piece of Victorian film can be taken in 1982 to any city on earth and screened. A newspaper printed in any language since roughly the same date will be clearly perceived to be a newspaper in any part of the globe, and many of its chief contents – puzzles, news, editorials, share prices, reviews – apprehended as such in scores of cultures where the language itself may not be known. Radio and television have developed more rapidly, but even with these, each new development – from the valve to the cathode ray tube, from color signals and transistors to cables – has required about fifteen years to become established within the market. Forms remain stable because the market keeps them so; the public's expectations of any particular device or genre take years to develop, and these expectations, transmuted into listener, viewer, or reader habits, are the capital assets of the publishers and companies that have discovered or nurtured them.

Yet behind the kinds of material and the hardware, important trends do make themselves apparent. Two that have been at work since the beginning of the century are worth emphasizing in any attempt to size up the changing information environment. One characteristic of the nineteenth century systems and devices – from the popular reading room to the peepshow – was that the audience was expected to make no investment in the system itself; revenue was derived either from the purchase of an artifact, such as a newspaper, or from the sale of a right, such as admission to a hall or tent. Indeed, the quest of the age had been so to multiply the product that the mass audience could have access at the lowest coin available. Thus arrived the half-penny newspaper, created as a result of expensive and diligent development of the mass press, mass distribution system, and the mass transportation system. As the century developed, however, the audience has been expected to indulge in an ever higher proportion of the total investment. Today, most of the investment necessary in maintaining a national television channel is held by the viewer rather than the supplier of the system – compare this with the theatre, or cinema, or the church.

In all of the new media, the audience's share of the investment has gone even higher, and the equipment companies have unsurprisingly been among the chief impresarios of development. The audience has to buy or rent the receiver and the recorder, the cable decoder, the videotex black box and so on. In fact, most of the new media are dependent upon there being several television receivers in a majority of homes; otherwise, there would be little hope of splitting the family as a viewing unit and thereby exploiting the potential for individual choice of material. With the arrival of direct broadcasting by satellite, the cost of each unit audience will rise substantially, since the engineering mechanisms required for switching from satellite to satellite, and thereby obtaining a wider choice, are fairly expensive. The whole expansion of the information sector thus hinges on the general expansion of the consumer economy, on the expansion at a steady rate of the consumer's propensity to invest in new entertainment systems. In the changeover from the old to the new systems, we are thus watching a very considerable switch in total investment in the resources of social communication from the manufacturer and the supplier to the audience at home.

The other important overall trend is for the gradual growth of local monopoly in any system. Information is historically torn between the condition of competition and its condition as a natural monopoly. One may cite the newspaper as a good example. Competition within the market for newspapers seemed natural, inevitable, and desirable in all democratic countries – as it still seems so today in places – so long as political circumstances made this desirable, and so long as the advertising done by mass consumer manufacturers required large slabs of display material. Gradually, television has become the channel for political material and the preeminent disseminator of national and regional manufacturer-to-consumer advertising. This has occurred on both sides of the Atlantic; although there are still a few European societies in which television advertising is illegal or minimal. Newspaper circulation has fallen a little in many countries, but seldom dramatically; where the total circulation has in fact fallen, the explanation often lies in the erosion of the habit of purchasing more than one newspaper as papers have become more comprehensive overall. The markets for advertising have, however, significantly altered throughout the economies of the West. The major area of growth has been in classified advertising, especially in recession-sensitive advertising such as that for jobs and contracts. The market has therefore become more volatile, while television advertising – dependent more upon manufacturers and sellers of commercial services – has tended to be much more resilient to temporary economic trends. There is a natural tendency for a newspaper to be most attractive as a source of advertising (and of news) where it is believed to be most comprehensive in its content, and this tendency, in the context of the changes mentioned, has greatly accelerated the development of natural local monopolies among the printed press. In the United States, this tendency has occurred alongside a growth of chains and of cross-media ownership at the corporate, if not the local, level. The newspaper has thus been coaxed by stages into becoming typically a local monopoly. Only a very few countries, such as Britain and Japan, have retained thoroughgoing newspaper competition, and in those cases, the reason has been the institution of national distribution, which has produced monopolization of another kind – within social strands rather than geographical location. Even in Britain, however, with its highly competitive journalism, the same phenomena have occurred with local newspapers as in Germany, the United States, France, and elsewhere.

The processes of monopoly have not set in as far as the electronic media are concerned, where cartelization is restrained through regulation. It would not be surprising, however, if a certain clarification did not begin – in those markets with a very large number of television outlets as the new media, with their far greater promise of abundant choice – to reach the middle and lower levels of the market. The same tendency toward a single outlet has occurred in the case of cinema, though mitigated by the habit of tripling or quadrupling movie theaters – not to create wider choice, but to provide finer tuning of the audiences for the existing repertoire as they grow and shrink during the run of a given film.

I have deliberately refrained from stressing national differences of trend or of magnitude throughout this paper, in order to bring out the shared phenomena of Western economics. We are witnessing a cultural shift, or set of shifts, that are more subtle and far-reaching than the physical devices, the products of modern electronics, themselves suggest, but that are more deeply rooted in the continuing and the slowly evolving than is often believed. After all, abundance of choice does not in itself constitute a transformation, since an individual will make conditioned choices and will probably not greatly increase the total hours of his exposure. But the role of text in our civilization and the development of the various skills of text are indeed in all probability today in the course of fundamental change. The management and use of a data base require quite new skills, and will emphasize different aptitudes from those required traditionally in primary education. The computerization of text suggests that we may absorb smaller quantities of text into our lives, but it will be text that is better ordered and more appropriately selected. The term *book* will probably come to cover a narrower range of products than it now does, and the technical aids to research will, in the late 1980s, enable a wider range of disciplines to benefit from the boon of the computer.

One might take, as an extreme example of the kind of "book" that is becoming outmoded, the telephone directory, where the form is used, in full Gutenbergian trappings of binding and single copy mass distribution, as the housing for a collection of data, only a tiny fraction of which is required by any individual reader. The time taken to collect the information and to reproduce it is so great that a high proportion of the material required by any individual reader is invalid by the time the finished product reaches him. As the total number of telephone subscribers rises, the proportion whose addresses change more frequently also rises. The directory is an essential body of data in urgent need of an appropriate mode in which to present itself to its readers. Clearly, the format of the traditional book is inappropriate, or will become so as soon as an alternative technology becomes as easily accessible, or where the level of accessibility of the alternative outweighs the disadvantages built into the existing mode. It is thus that the book will "die," not through sudden technological redundance, but through the prudent choices of those who actually require the information it carries. As the newspaper passes through its own crises of form, many of its traditional elements will probably be lost to the new electronic mode. The pursuit of information "abundance" in this case is in reality the pursuit of a manageable modicum of relevant information.

But for the most part, the contemporary drive for abundance of choice is a besetting ideology much more than a practical need. It is more like the Victorian illusion of mechanized immortality, providing evidence for the psychic tension of the moment rather than for a social or economic need. The pursuit of plenty in the sphere of information is a psychological analogy to its pursuit in the sphere of nourishment in the developed world, where the use of food has more to do with marketing, fashion, and general culture than with biological need. Information abundance has likewise much to do with cultural identity in the late twentieth century, and little to do with need. Nonetheless, it is a motive force and a justification for an industrial evolution with revolutionary repercussions. As a recent OECD report observes:

> The production, transmission and processing of the most varied information will be at the heart of economic activity and social life ... through its links with data processing and telecommunications, the electronics complex during the next quarter of a century will be the main pole around which the productive structures of the advanced industrial societies will be reorganized.

There are plenty of documents in circulation that outline the growing disparity in the provision of information, and especially the communications technology, between the countries of the North and those of the developing South. Eighty-five per cent of existing data bases are in the North; 70 per cent are in the United States. One company, IBM, is responsible for manufacturing two thirds of the world's computers. The abundance both of hardware and software is the privilege of a tiny group of societies, who are themselves enjoying a continuingly increasing disparity. Information grows by what it feeds upon.

Studies of the flows of data from computer to computer reveal its increasing internationalization. The growth of international networks is growing in the wake of the establishment of effective national networks: EURONET is establishing itself after the American TELENET and TYMNET systems. Even the countries of the socialist camp are now becoming connected to Western economics through the flows of international data. There is a strong tendency for all data flow to be concentrated on capital cities, however, because that is where the main data users are located. But there is also a tendency for data to flow from the less developed to the more developed, where processing facilities are more plentiful and more efficient. In the three least advanced countries of Europe, for example, over a quarter of their national data flow is toward other countries. The revenues from computer services in the United States have more than quadrupled in the last decade ($2 to $8+ billion), the international element tripling ($300 to $900 million). The newly emerging techniques of remote sensing and satellite distribution of data are bringing about further exponential growth and further tiltings in the international flow of data from developing to developed worlds. The cultural implications are self-evident, and the political implications, easy to deduce. Behind the emblem of information foison there exists a growing phenomenon of global cultural domination, produced not by powerful armies, but by powerful international companies. The greater the stock of expertise in a society, the greater is its ability to make use of the information technology and benefit from its software. The educated society is the one best suited to prosper in the new age, and everything conspires against the society that has a deficit in its national balance of educated talent. The profusion of data through which the Western industrialized consumer indulges his or her choice, and expresses the nuances of a carefully refined and nurtured life-style, is the same oversupply that is drawn from the international data flow and jeopardizes the nationhood of developing societies. We may expect in the next decades the lines of international tension to shadow the contours of data abundance.

Knowledge production and occupational structure

Fritz Machlup

The focus of our attention will now be on the worker, his activity and occupation, rather than on the industry in which he is employed. The question is whether, how, and to what extent the occupational composition of the labor force and of employment has changed, and how this change is connected with the changing role of knowledge-production in the economy.

Technology, demand, and occupational structure

It may seem "self-evident" that great advances in technology and great shifts in demand lead to changes in the occupational composition of the labor force – but actually it is neither self-evident nor necessary in any sense. It is conceivable that all sorts of technological progress and shifts in demand leave the occupational structure of the economy unchanged, provided occupations are not too narrowly defined.

Inevitable, impossible and probable changes

Assume that a technological revolution in an industry allows the same output to be produced with half the quantity of physical labor, or twice the output with an unchanged work force. Even if demand is not sufficiently elastic to permit the same work force to be kept on to produce the doubled output, the workers released from this industry may find jobs requiring very similar kinds of work in other industries. If they cannot find such jobs at their accustomed wage rates, they may find them at relatively reduced wage rates.

Assume next that a drastic switch in demand occurs which, at given price relations, would call for more output of a product made with one type of labor, and less output of a product made with another type of labor. If workers are not flexible and cannot change over from one occupation to another, wage rates will rise in the one industry and fall in the other, and the prices of the products in question will do likewise, until the quantities demanded, despite the changes in demand, will be adjusted to the supply "dictated" by the inflexible occupational structure of the labor force.

Thus neither changes in technology nor in demand *necessarily* result in changes in the composition of the labor force. With inflexible skills and preferences on the part of labor, and flexible relative wage rates, the employed labor force can remain unchanged in composition. Its composition will change only where the labor force is adaptable and adjusts to the "requirements" of changed technology or changed demand.

In the United States, the labor force has been adaptable to a high degree. Workers, by and large, follow the monetary incentives created by changes in relative earnings; even where the wage rate is not sufficiently flexible, they respond to the stimuli of open job opportunities. For this reason it is highly probable that changes in technology or demand will result, with a natural time lag, in a corresponding change in the occupational structure of the labor force. (Incidentally, there can be considerable adaptability in the long run even without any workers changing their occupations: all that is needed is that new entrants into the labor force go into the occupations now favored by the changes in technology or demand, and allow other occupations to decline through death and retirement.)

Even with a highly adaptable labor force, it is not certain that every technological change releasing physical labor in a given industry will necessarily result in a decline in the ratio of physical labor to the labor force as a whole. Consumers, having more of their incomes left to spend on other things when one product becomes cheaper owing to improved technology, may well choose to buy goods and services produced with an even greater share of labor of the type displaced – so that no change in the composition of the labor force may be called for. But to say that not every technological change needs to have an effect upon the occupational structure is one thing; it is another to say what is likely to happen when many technological changes occur. If physical labor is displaced in a large number of industries, the probability is great (1) that some of the consumers' purchasing power will be used for things made with less physical labor and more of other types of labor, (2) that, on balance, therefore, demand will change in favor of these other types of labor, and (3) that consequently there will be a change in the composition of the labor force such that the share of physical labor is reduced.

It helps in the interpretation of observed statistical trends if one clearly understands the difference between what is logically necessary and what is logically probable. The "logic" in this connection refers to the concepts formed and their interrelations posited within the explanatory models of the economy.

Mechanization and automation

Much of the technical progress of the past has consisted of the replacement of men by machines in particular production processes. This will undoubtedly be true also for the technical progress of the future. In recent years much has been heard about "automation" and how it threatens to depopulate entire industries. Automation and mechanization are often confused with each other, though it should not be too hard to keep them apart. Mechanization saves the use of human muscles, automation saves the use of human judgment. Mechanization, therefore, displaces physical labor, while automation displaces mental labor. These displacements occur in particular processes, not necessarily in the economy as a whole. The reabsorption of the displaced labor in other processes, industries, or sectors of the economy, may or may not be causally related to the displacement.

It is not possible to draw a clear and incontrovertible line between "physical" and "mental" labor. Almost every kind of operation requires both physical and mental effort. Even pressing a button, pushing a pen over a piece of paper, or dictating an order call for some muscular activity; on the other hand, lifting a bag, shoveling snow, carrying loads cannot be done without some mental activity. Yet, for theoretical as well as practical purposes it is possible to make the distinction between physical and mental operations, and between predominantly physical and predominantly mental labor. The frequently-used distinction between blue-collar and white-collar workers is designed precisely for this purpose.

When an "automatic control" device reaches "conclusions" and makes "decisions" which even the best human brains could not achieve with the same accuracy and/or the same speed, things become possible which brain power alone simply could not do. In these cases, automation does not "replace" mental labor hitherto employed, but merely permits things to be done that had not been possible before; the "thousands" of brains that are replaced in such instances are only hypothetical brains hypothetically employed. In other cases, however, there is actual replacement of human labor as a result of automation. Masses of clerical workers may be replaced by automatic bookkeeping and data-processing machines. In such instances, automation may reduce the share of white-collar labor in the work force of particular firms or industries.

In particular firms and industries mechanization may reduce the share of physical labor, while automation may reduce the share of mental (clerical, administrative) labor employed. This does not mean that automation tends to reduce the share of brain workers in the labor force of the economy as a whole. The design of automatic devices and their introduction and operation require brain workers of a higher calibre. This suggests that the distinction between physical and mental labor is really not adequate for the description of the consequences of labor replacement by machines. What we would need is a stratification of both physical and mental labor: physical labor of various degrees of manual skill, and mental labor of various degrees of analytical skill. On balance, the end effects of both mechanization and automation may then be found to lie in the demand for more highly skilled labor, manual and mental.

It would go too far beyond the scope of this volume to test this hypothesis on the basis of the statistical data at our disposal, though the census data on the occupational distribution of the labor force would be detailed enough to attempt a stratification of the sort proposed. At this juncture, we shall confine ourselves to two tasks: first, to observe what changes have occurred over the last 60 years in the relative distribution between "white-collar," "manual and service," and "farm" labor; and, secondly, to view the changing picture, over the same period, involving two groups, "knowledge-producing" and "not knowledge-producing" or "primarily manual" labor. (These findings will be interesting enough, but I do intend in the near future to study the changing distribution of the labor force over *several* strata of skill in both knowledge-producing and manual occupations.)

The changes in the occupational structure

The U.S. Bureau of the Census publishes occupation statistics for all "employed" and for all "economically active" persons. "Employed" includes "self-employed"; "economically active" includes unemployed with previous work experience (who are listed in the occupation in which they were last employed). The labor force is divided into almost 400 occupations within eleven "major occupation groups." These eleven groups are: (1) Professional, technical, and kindred workers, (2) farmers and farm managers, (3) managers, officials, and proprietors, except farm, (4) clerical and kindred workers, (5) sales workers, (6) craftsmen, foremen, and kindred workers, (7) operatives and kindred workers, (8) private household workers, (9) service workers, except private household, (10) farm laborers and foremen, and (11) laborers, except farm and mine. The data are obtained from the decennial population-census *enumerations* and from quarterly *sample surveys.* In the tables used in the following discussion, the figures for 1900, 1910, 1920, 1930, 1940, and 1950 come from the decennial population census, the figures for 1959 from the current population survey.

White-collar, manual and service, and farm workers

The eleven "major occupation groups" can be regrouped into three categories: (A) "white-collar workers" consisting of groups (1) professional, technical, etc., (3) managers, officials, and proprietors (except farm), (4) clerical, etc., and (5) sales workers; (B) "manual and service workers" consisting of (6) craftsmen, foremen, etc., (7) operatives, etc., (8) private household workers, (9) service workers (except household) , and (11) laborers; and (C) "farm workers" consisting of (2) farmers and farm managers, and (10) farm laborers and foremen. Table X-1 shows the number of persons in each of these three categories for given years from 1900 to 1959. This table shows that the number of white-collar workers increased steadily from 5 million in 1900 to 27 million in 1959, the number of manual and service workers increased from 13 million to 31 million, and the number of farm workers decreased from 11 million to 6 million. Thus, in 1959, white-collar workers were 540 per cent, manual and service workers 238 per cent, and farm workers 59 per cent of their respective numbers in 1900. In the 19 years between 1940 and 1959 the number of white-collar workers increased by 69 per cent, and the number of manual and service workers by only 16 per cent, while the number of farm workers decreased by almost 29 per cent.

The distribution of the labor force among the three categories is brought out in stronger relief if it is shown as a percentage of the total. This is done in Table X-2. Farm workers, 37.5 per cent of the labor force in 1900, were only 9.9 per cent in 1959. Despite this decline in the percentage of farm workers, other physical workers, manual and service, increased only from 45 to 48 per cent of the total labor force.

TABLE X-1

Labor Force, or Economically Active Civilian Population, by Broad Occupation Categories, 1900–1959 (millions of persons)

Category	1900	1910	1920	1930	1940	1950	1959
White-collar	5	8	11	14	16	22	27
Manual and service	13	18	20	24	27	30	31
Farm	11	12	11	10	9	7	6
Total	29	37	42	49	52	59	65

TABLE X-2

Labor Force: Percentage Distribution Over Broad Occupation Categories, 1900–1959

Category	1900	1910	1920	1930	1940	1950	1959
White-collar	17.6	21.3	24.9	29.4	31.1	36.6	42.1
Manual and service	44.9	47.7	48.1	49.4	51.5	51.6	48.0
Farm	37.5	30.9	27.0	21.2	17.4	11.8	9.9
Total	100.0	100.0	100.0	100.0	100.0	100.0	100.0

If all manual workers, industrial and agricultural, are taken together, their combined share in the labor force decreased from 82.4 per cent in 1900 to 57.9 per cent in 1959. White-collar workers, conversely, increased from 17.6 per cent of the labor force in 1900 to 42.1 per cent in 1959. This trend, uninterrupted for 60 years and probably longer, is most impressive.

Knowledge-producing and not knowledge-producing workers

The "major occupation groups" distinguished by the Census Bureau, and the three occupation categories into which they were regrouped, are rather rough classifications for tests of the kinds of hypotheses that have suggested themselves in the course of the discussions in this survey of knowledge-production. Can all white-collar workers be regarded as knowledge-producing workers? Surely not without several qualifications or adjustments. In order to obtain a distribution of the labor force more suitable for our purposes, we must examine all occupations in the major groups distinguished by the Census Bureau, and make appropriate exclusions and inclusions.

[...] Transporters, transformers, processors, interpreters, analyzers, and original creators of communications of all sorts will be regarded as knowledge-*producing* workers. (A narrower definition might well have been more appropriate, but we shall not change it at this point.) On the other hand, knowledge-*using* workers, however knowledgeable, will not be included if their product is not a communication or a service contributing to knowledge-transmission. If their product is a message, a piece of information, anything primarily designed to create an impression on someone's mind, they will be included among knowledge producers, even if their own mental equipment is relatively poor. (This is emphasized as a reminder and a warning, not as a contention that this is necessarily the best thing to do.) Under this set of rules, the insurance salesman and the mail clerk are in the class of knowledge producers, but dentists and veterinarians are not.

The following adjustments are made to carry out these resolutions. From the group of "professional, technical, and kindred workers," the first in the white-collar category, the following occupations are excluded as not knowledge-producing: chiropractors, dentists, funeral directors and embalmers, all nurses, pharmacists, 50 per cent of physicians and surgeons (on the assumption that only one half of their work is diagnostic and therapeutic advice and prescription), all technicians (medical, dental, testing, etc.), therapists and healers, and veterinarians. From the group "managers, officials, and proprietors (except farm)", railroad conductors, and all salaried or self-employed managers in retail trades, automobile repair services and garages, and miscellaneous repair services are excluded on the grounds that they usually do not confine their activities to managerial tasks but participate in the physical work carried on in their shops. From the group "sales workers," hucksters, peddlers, and all persons in retail trade are excluded because they are less specialized in "sales talk" than in handling the merchandise sold. Finally, from the group "craftsmen, foremen, and kindred workers," the first in the blue-collar category, the members of the printing trades, particularly the electrotypers and stereotypers, engravers, photoengravers and lithographers, compositors and typesetters, pressmen and plate printers are shifted into the class of knowledge-producing workers.

These adjustments cause no difficulties in the census years. For 1959, however, the absence of detailed breakdowns necessitates some auxiliary manipulation. Assuming that the various detailed occupations had in 1959 the same relative membership, in per cent of the "major occupation group" to which they belong, as in 1950, we use the 1950 shares of group totals to find the 1959 numbers of workers to be "transferred" from the knowledge-producing class to the not knowledge-producing class, or *vice versa*. Another makeshift is necessary for all census years because the numbers of retail sales workers are not separately stated. (They are given for 1959.) To make an educated guess of the numbers of retail sales workers, let us assume that they are equal to the numbers of managers or proprietors in the retail trades, an assumption suggested by the fact that they seem to consist chiefly of small-scale establishments with one hired worker helping the owner. The available figure for 1959 indicates that this procedure involves only a slight underestimation of the numbers of sales workers in retail trade in the preceding census years, say 1940 and 1950. [...]

The division of the civilian labor force into Classes I and II shows a steady increase of Class I and steady decline of Class II, the not knowledge-producing workers. Class I, the knowledge-producing occupations, increased from 10.7 per cent of the labor force in 1900 to 31.6 per cent of the labor force in 1959. Class III, the potential members of the labor force who worked on producing knowledge in their own heads by going to school, increased from 3.2 per cent in 1900 to 19.4 per cent in 1959. These are percentages of the sums of Classes I and II. Expressed as percentages of the sums of all three classes – that is, of the potential civilian labor force – the combined membership in both kinds of knowledge-producing occupations, Classes I and III, increased from 13.5 per cent in 1900 to 42.8 per cent in 1959. The decision to combine Classes I and III makes good sense or not, depending on the purposes of the analysis. If the purpose is to show how society has changed its allocation of human resources among alternative activities, one will want to see the knowledge-transmitters and the full-time knowledge-receivers combined. If the purpose is to show how the active labor force has changed its composition, largely in response to technological progress and economic growth, one will confine his attention to the relative role of the knowledge-transmitters.

Different growth rates within the class

Since the class of knowledge-producing occupations consists of five major occupation groups, we ought to examine whether the large increase is equally distributed among the groups or whether one or another of these groups is chiefly responsible for it. We shall find that over different periods different occupation groups contributed most to the growth of the class.

Looking first at the change in composition over the entire 60-year period, we find that the clerical occupations grew most remarkably, from 3 per cent of the labor force in 1900 to 14 per cent in 1959. But the other occupations in the class of knowledge-producers also assumed enlarged shares in the labor force, except the printers, who barely maintained theirs. Professional and technical workers grew from 3.6 per cent in 1900 to 8.4 per cent in 1959; managers, officials, and proprietors from 2.5 per cent to almost 6 per cent; sales workers (outside the retail field) from 1 per cent to 2.7 per cent.

Taking only the most recent decade into account, the improvement in the relative position of the professional-technical group is the most outstanding: from 6.8 per cent in 1950 to 8.4 per cent in 1959. This may indicate a strong impact of the technological "revolution" of which engineers and operations researchers have been telling us. On the other hand, the decline shown in the share of the sales workers (except retail) must not yet be accepted as a fact, because it may be merely the reflection of the probable underestimation of the retail sales force in 1940 and 1950 that was mentioned above.

[...] Over the period from 1900 to 1959, when the total civilian labor force increased by 137 per cent, all knowledge-producing occupations increased by 602 per cent, with the clerical workers showing the fastest growth rate: 1,002 per cent. Over the 19 years from 1940 to 1959, the labor force increased by 23 per cent, all knowledge-producing occupations by 80 per cent, while the group of managers, officials, and proprietors led with a growth of 103 per cent. Over the last nine years, from 1950 to 1959, the increase in the labor force was 17 per cent, the increase in all knowledge-producing occupations 30 per cent, and the group of professional and technical workers grew at the fastest rate, 43 per cent.

The income shares of knowledge-producing occupations

If the remarkable rise of knowledge-producing occupations in the total labor force is interpreted as an "adjustment" to other economic changes and as a "response" to economic incentives, it would be interesting to see whether the change in the occupational structure was associated with an equal or greater change in income distribution between the two broad occupation classes. If the income share of the knowledge-producers has increased at a faster rate than their share in the employed labor force, one might interpret this as supporting the hypothesis that earnings differentials played a significant role in the change in the occupational structure. If, on the other hand, the income share increased only at the same rate or more slowly, or not at all, other hypotheses would be suggested; for example, that the mere availability of job opportunities accomplished the relative reallocation usually attributed to differentials in the rates of earnings; or that the supply of knowledge-producing labor had increased faster than the demand, causing the relative earnings of such labor to decline and its income share to lag behind its increased share in employment. In any case, we should like to see what happened.

The incomes in knowledge-producing occupations

In order to be able to assign income data to occupational groups, we shall have to use (1) the "gainfully employed" labor force rather than the "economically active" labor force (which includes unemployed); and (2) separate figures for male and female workers, since their earnings differ substantially (not so much for "equal work" as because they are given different kinds of jobs).

Census data are available for median wage-or-salary incomes for both sexes in each major occupation group. To estimate aggregate incomes we should appropriately have arithmetic means rather than medians, but the means have been estimated only for two years, 1939 and 1949, which is not sufficient for our purposes. In using the medians as a basis for computing the total incomes earned by various occupation groups, we shall have to bear in mind the likelihood of substantial underestimations, and in computing the relative income shares, the possibility of distortions because of different deviations of the medians from the arithmetic means. On the other hand, since we are interested only in changes over time, it is not at all certain that the arithmetic means, if they were available, would give us more relevant information. For the relative attractiveness of different occupations the median incomes may surely be pertinent; and while it may not be sensible to multiply the number of persons by their median income in order to obtain total group incomes, it is sensible to do this multiplication in order to get weighted averages of the medians for aggregations of occupation groups.

[...]

We find that in 1940 a total of 10,573,000 knowledge-producing workers were employed and earned a total of $15,120 million. In 1950 the number of this type of workers employed was 15,341,000 and their income $44,783 million. In 1958 there were 20,497,000 employed, earning $89,960 million. Their average annual incomes were $1,430 in 1940, $2,919 in 1950, and $4,389 in 1958 – increases reflecting in a large part the inflation of the period. To interpret the changes in the total incomes of knowledge-producing workers we may compare these incomes (1) with total "employee compensation," (2) with the sum of employees' and proprietors' incomes, or (3) with national income. Since proprietors are included in one of the occupation groups, their incomes ought to be included too, though these incomes probably contain some returns other than labor income.[...]

The shares in total income

The results of our calculations, reproduced in Table X-3, conform with our expectations regarding the changes from 1950 to 1959, but are unexpected so far as 1940–1950 is concerned. Despite the large increase in the share of knowledge-producing workers in the total labor force and the similar increase in their share in total employment, their share in total income was the same in 1950 as in 1940. The absence of any growth in their income share is apparent in all three ratios, as per cent of total employee compensation (line 5), as per cent of employees' and proprietors' incomes (line 6), and as per cent of national income (line 7). We must remember, of course, that these "findings" may be only statistical illusions, created by clumsy or crafty juggling of data; and that these data in turn are more or less reliable estimates by economic statisticians who have learned that poor figures are preferable to none. But assuming that the findings conform to actual fact, that is, assuming that between 1940 and 1950 the average salary or wage of knowledge-producing workers has actually declined relative to other wages, how can this fact be explained?

Two hypotheses were advanced for such an event in the introduction to this section. One posits an increase in job opportunities for members of knowledge-producing occupations at a time when job opportunities for

TABLE X-3

Relative Employment and Relative Incomes from Employment in Knowledge-Producing Occupation Groups, 1940, 1950, and 1958

		1940 (incomes for 1939)	1950 (incomes for 1949)	1958 (incomes for 1958)
(1)	Total employed persons	44.9 million	56.2 million	65.6 million
(2)	Knowledge-producing persons employed	10.6 million	15.3 million	20.5 million
(3)	Share in total employment $\left[\frac{(2)}{(1)} \times 100\right]$	23.6%	27.3%	31.3%
(4)	Income of all knowledge-producing occupations	\$15,120 million	\$44,783 million	\$89,960 million
(5)	Income of all employees	\$52,129 million	\$154,190 million	\$277,400 million
(6)	Income of all employees and proprietors	\$65,139 million	\$191,731 million	\$303,386 million
(7)	National income	\$81,634 million	\$241,876 million	\$336,183 million
(8)	Share in employees' income $\left[\frac{(4)}{(5)} \times 100\right]$	29.0%	29.0%	32.4%
(9)	Share in income of employees and proprietors $\left[\frac{(4)}{(6)} \times 100\right]$	23.2%	23.4%	29.7%
(10)	Share in national income $\left[\frac{(4)}{(7)} \times 100\right]$	18.5%	18.5%	26.8%

other types of labor are limited. In such a situation there can be an influx into the occupations favored by better chances of employment even if they do not promise chances for higher earnings than in other occupations. An alternative hypothesis posits an autonomous increase in the supply of workers in knowledge-producing occupations which is not matched by an autonomous increase in demand. A third hypothesis can be based on a peculiarity of periods of inflation, namely, differences in the speed with which incomes in different occupations are adjusted to expansions of effective demand. It is quite likely that wages and salaries in knowledge-producing occupations are less flexible than wages of manual labor, and thus rise more slowly during an inflationary period such as that between

1940 and 1950. Closer inspection [...] however, reveals a fact which is almost sufficient to explain the relative decline of the earnings rates of knowledge-producing labor from 1940 to 1950: the number of women in clerical work increased by almost 80 per cent and this group commands the lowest pay of all workers in knowledge-producing occupations. (The wage of female sales workers is still lower, but they were largely excluded from Class I.)

From 1950 to 1958 the share of knowledge-producing occupations in total income increased substantially; the income of this class of workers rose from 29 to 32.4 per cent of total employee compensation; from 23.4 to 29.7 per cent of the sum of employees' and proprietors' incomes; and from 18.5 to 26.8 per cent of national income. These increases match or exceed the relative increase in employment in knowledge-producing occupations, which suggests that the demand for labor in these occupations increased more than the demand for other types of labor.

Different changes of income shares within the class

We strongly suspect that our broad classification involves aggregations of occupation groups so heterogeneous that the most significant structural changes remain hidden. Although professional, technical, clerical, and sales workers all earn their living chiefly by talking and writing – and are for this reason thrown together as members of knowledge-producing occupations – the kinds of talk and the kinds of pen-pushing and paper-shuffling are sufficiently different to merit separate treatment in many respects, particularly with regard to their employment and earnings opportunities. As a matter of fact, the statistical data[...], however crude, do show that the income shares of particular occupation groups within the class of knowledge-producers have not all changed the same way as the share of the entire class.

Confining ourselves to one of the three ratios, that of the occupational-group income to the sum of all employees' and proprietors' incomes, we find that during the war-and-inflation decade, 1940–1950, the "managers, officials, and proprietors" and the clerical workers lost some ground. The income share of the former fell from 6.1 to 5.9 per cent, the share of the latter from 8.7 to 8.5 per cent. The sales workers barely held their ground, as did the printers, despite their strong trade union. On the other hand, the group of professional and technical workers – although it includes the teachers, who were notorious losers in this respect – improved its income share from 5.9 to 6.5 per cent. Thus, the high-level-knowledge-producers – if we may assume that this group contains more knowledge interpreters, analysts, and creators than the others – gained both from 1940 to 1950 and from 1950 to 1958.[...]

Some implications

Despite the poverty of the statistical series they do appear to establish certain facts and strong trends that are significant for our discussion.

The facts and the trends

Before we formulate what can be said about trends exhibited by our data, we should refer to one or two facts in addition to those already cited, for they throw a sharp light upon the employment situation of manual and white-collar workers.

From 1950 to 1959, while the total number employed increased, employment of "production workers" in manufacturing declined both relatively and absolutely; the employment of nonproduction workers increased by over 48 per cent. This rise in employment of nonmanual labor and fall in employment of manual labor are also reflected in the incidence of unemployment: the ratio of unemployed to employed is now twice as high among blue-collar workers as among white-collar workers.

These are the trends read from the statistical series: (1) The knowledge-producing occupations have grown over the last 60 years much faster than occupations requiring manual labor. (2) The share of knowledge-producing occupations in the total labor force tripled between 1900 and 1959. (3) The share of these occupations in total employment has increased even more. (4) While in the first part of this century growth was fastest in clerical occupations, the lead was then taken by managerial and executive occupations, and more recently by professional and technical personnel. (5) The share of knowledge-producing occupations in total income has increased during the last decade. (6) The share of professional and technical personnel in total income has increased during the last two decades.

Interpretation

From these trends one may safely conclude that both the supply of and the demand for knowledge-producing personnel have been increasing. In particular it seems safe to conclude that technological progress has been such as to favor the employment of knowledge-producing labor; that shifts of demand for final products have been such as to increase the demand for such labor over the demand for unskilled manual labor, and that the supply of labor suitable for knowledge producing occupations has increased, presumably as a result of more widespread school attendance.

Heavy emphasis, I submit, should be placed upon the fourth of the trend-propositions formulated above. It asserts a trend within a trend: while the ascendancy of knowledge-producing occupations has been an uninterrupted process, there has been a succession of occupations leading this movement, first clerical, then administrative and managerial, and now professional and technical personnel. Thus, the changing employment pattern indicates a continuing movement from manual to mental, and from less to more highly trained labor.

Implications

If employment opportunities continue to improve for high-level-knowledge-producing labor and to worsen for unskilled manual labor, the danger of increasing unemployment among the latter becomes more serious. To speak of absolute unemployability of people of low intelligence and little training may be going too far, because employability is partly a matter of the price at which labor is offered. But since society no longer tolerates "cheap" labor, and unskilled physical labor may find uses only if it is cheap, the combination of our social ideas with the continuing technological and economic trends may in fact spell unemployability for certain low-level types of labor. At socially acceptable wage rates, workers of very low economic productivity may remain permanently unemployed; and this unemployment is apt to persist even in the face of attempts to create "effective demand" if wage rates are promptly adjusted to inflated price levels.

A cogent exposition of this "Theory of Creeping Unemployment and Labor Force Displacement" has been presented by Clarence D. Long. The continuing upward creep of the unemployment rate in prosperity periods is, according to this theory, the result of a widening productivity spread and a constant relative-wage spread. Productivity of above-average workers increases faster than productivity of below average workers: thus productivity differentials among different types of labor become greater. If average wage rates rise at the rate at which average productivity increases, and if the lowest wage rates rise at the same rate as the average – so that the percentage spread between minimum and average wages does not increase – it follows that the minimum wage rates rise faster than the productivity of low-level labor, with the result that such labor can no longer be employed with profit. (The minimum wage in this context need not be that prescribed by law; custom or employer ethics may forbid paying less.)

Long's theory is sufficiently general to apply to any kind of labor offered by workers of different ability and training. It applies, however, with special relevance to the situation described here, in which employment opportunities for low-skilled physical labor have been declining, and the economic productivity of high-level-knowledge-producing labor increasing. The implications, then, of the trends observed with regard to the occupational composition of the employed labor force are rather dismal. They seem to leave us with an unpleasant choice: either to resign ourselves to larger wage differentials, increasing spreads between minimum and average earnings, or to face a continuing upward creep of the rate of unemployment, not only in bad times but also in prosperity. Perhaps this dilemma can be avoided by a third possibility, namely, through a drastic improvement of school programs that raises the lazy and unambitious to higher levels of accomplishment. But even if this is a possibility, it can be realized only years after the school reform, a reform which probably is not much less unpopular than low wages or unemployment.

Industries and occupations: ratios and growth rates

One more promissory note remains to be redeemed. We have promised to compare the findings of the occupation approach with those of the industry approach. In particular, we want to compare (1) the total product of, or rather expenditures in, knowledge-producing *industries*, relative to the total product of the economy, with the total income of knowledge-producing *occupations*, relative to the total income of the economy; and (2) the growth of the output of or expenditures in knowledge-producing *industries* in the recent past with the growth of income of knowledge-producing *occupations* during approximately the same period.

Ratios to total product

The first of these comparisons requires only that we set side by side ratios computed earlier in this chapter and in the preceding one.

We listed the 1958 expenditures in all branches of knowledge-production and obtained a total of $136,436 million, of which $109,204 million were regarded as "final product." Since this contained several items not included in the official Gross National Product, we added these items to the GNP and obtained approximately $478,000 million as the adjusted GNP for 1958. The expenditures for knowledge-production were almost 29 per cent of the adjusted GNP. The expenditures for knowledge as final product were 23 per cent of the adjusted GNP.

We tabulated estimates of the incomes of workers in knowledge-producing occupations and obtained for 1958 a total of $89,960 million. The national income of the same year was $336,183 million. As shown in Table X-3, the income of knowledge-producing workers was 26.8 per cent of the national income in 1958. Neither the knowledge-workers' income nor the national income contained the earnings foregone by persons engaged in education. If these potential earnings, which we had included among the cost of education – $13,519 million for high-school students, $7,189 million for students in higher education, and $4,432 million for mothers staying home to educate their pre-school children, together $25,140 million – are added, we obtain $115,100 million as the total of actual and foregone incomes of knowledge-producing workers, and $361,323 million as the national income adjusted for the incomes foregone by potential workers giving or receiving education. The former figure is 32 per cent of the latter.

This, then, is the set of ratios: 29 and 23 per cent by the industry approach, and 27 and 32 per cent by the occupation approach. These ratios refer to different things and, naturally, must not be expected to be the same. The discrepancies are certainly not greater than one should expect from an understanding of the concepts involved. As a matter of fact, there is no reason why the various approaches should yield ratios as similar as the above ones. When we first considered the possibility of the two approaches, we pictured an imaginary economy so organized that the results of the two approaches would be almost the same; and we concluded that the economies of the real world were quite different from the imaginary one.

Growth rates

There is some reason for expecting that "growth rates" derived from the two approaches are similar, though any optimism in this respect should be tempered by a realization of the roughness of many of the estimates by which our statistical data were obtained.

The industry approach yielded "growth rates" for the four-year period 1954–1958 and for the eleven-year period 1947–1958. The occupation approach furnished growth figures for the eight-year period 1950–1958, for the 18–year period 1940–1958, and for even longer intervals. The only comparisons that can reasonably be made are between annual growth rates computed from the eleven-year period of industry growth and from the eight-year period of growth of workers' incomes.

The weighted average of the approximately eleven-year growth rates of 36 different knowledge industries [...] was 10.6 per cent per year.

For all workers in knowledge-producing occupations the incomes from employment rose from $44,783 million in 1950 to $89,960 million in 1958, or by 100.88 per cent. This growth over eight years constitutes a growth rate of 9.1 per cent per year.

The similarity of these results, 10.6 and 9.1 per cent, is close. No attempt has been made to reconsider or revise any of the estimates in the hope of achieving greater correspondence in the results from the two approaches.[...]

The reliability of the data with which we worked must not be overestimated, and the legitimacy of several of the uses we made of them may be questioned. Indeed some of the statistical procedures were accepted only as makeshifts and in the hope that others may improve upon our most imperfect effort.

Statistical accuracy and general conclusions

Our attitude of caution regarding the accuracy of statistical data and the reliability of ratios between them should not be misread to imply serious reservations concerning the validity of the generalizations [...]. Many things can be said about general trends and interrelated developments even if the illustrative or supporting data are less than accurate. What has been said about changes in the occupational structure remains valid even if the basic statistics are somewhat off the mark. And what has been said about the cost and efficiency of education, or research and development, is not vitiated by flaws that can be detected in the statistical data. Thus I hope no one will discard the proposal for educational reform just on the grounds that some of the estimates of the expenditures or implicit costs may be questionable.[...]

Information and society

Edwin B. Parker

There are three basic elements that make up our new, largely man-made environment and, hence, determine the quality of life. The first is matter, the second is energy, and the third is information.

The three factors are related. The supply of matter and energy on this planet is finite; consumption of matter and energy cannot increase indefinitely without endangering the survival of the species. But information, which is the pattern of organization of matter and energy, has a potentially infinite supply. All our increases in material wealth flow from the discovery or the creation of new patterns in our matter and energy. Investment in the production of information (e.g., through research and development) and investment in widespread distribution of information (e.g., through education) may be the only way to permit continued improvement in the quality of life without large increases in consumption of matter and energy.

All of the society's expenditure on science and technology and on research, development, and discovery in all fields can be viewed as investment in the production of information. Policies concerning resource allocation for scientific and technical information must be considered in the context of that larger question of science policy generally. There are several related questions that need satisfactory answers. How much of society's resources should be allocated to research and development? How much of the research and development funds should be allocated to supporting scientific and technical information services? How can those funds be most efficiently used? How should funds be allocated between provision of present information services and research and development leading to more efficient future services?

Similarly, all of society's expenditures on education, broadly defined, can be viewed as investment in the distribution of information. Policies concerning resource allocation for school and public libraries must be considered in the context of that larger question of education policy. How large a share of society's resources should be allocated to education and other channels for the distribution of information? Can we afford to make education available to everyone throughout his lifetime on an equal opportunity basis? To what extent can education be made more productive (e.g., less costly without reducing quality), through the use of books and other media as a substitute for expensive human labor? How many of our education dollars should go directly into education services now, and how many should go into the development of media and institutions that can make future education less costly or more widely accessible?

Our society spends much money on the telephone network, communication satellites, computer information systems, broadcast transmitters, cable television, printing presses, photocopy machines, library buildings, microfiche readers, record players, audio and video cassettes, and other information utilities. How should we allocate resources between the technology for the storage and distribution of information and the preparation of messages to be stored in and transmitted through that technology? Can investment in improved information technology significantly reduce the unit costs of storing and transmitting information? Could an improved infrastructure of information technology permit the substitution of information for matter and energy – for example, by replacing some human travel with electronic teleconferencing? Does our economic system provide sufficient incentive for the development and installation of a technological information infrastructure that will lead to productivity gains in both the private sector (e.g., in business and industry) and the public sector (e.g., in science and in education) of the economy? Are there major economies of scale to be gained such that public-sector information services would be cheaper as part of a general-purpose information network (e.g., via communication satellite and cable television) than if dedicated, special purpose networks were developed?

The role of information in society will be determined by how our society answers these economic questions concerning the production of information, the distribution of information, and the technological infrastructure for information storage and transmission. Whichever way the questions are asked, the social effects will follow from the decisions allocating the economic resources of the society. The fact that they are asked as economic questions may be disturbing to those who are now questioning the value of economic growth as a social goal. Many have begun to question the value of continued expansion in our capability to produce material goods. The shift in emphasis from material goods to services has already led to our society being called "post-industrial." As the following analysis attempts to show, the fact behind that label is a relative decline in the percentage of our economy devoted to matter and energy and a relative increase in the percentage of the economy devoted to information. If we are to further shift our national economic goals from the narrow perspective of production and distribution of matter and energy, and focus instead on the quality of life and maximizing human potential, then even further shifts in allocation of economic resources to information may be called for.

John Diebold, writing in the April 1973 issue of *Foreign Affairs*, argues that we are now beginning the third phase of the 200–year-old industrial revolution. The third phase (or third century) of that revolution will be increasingly concerned with production and distribution of information, he says. He argues that the first century of the industrial revolution was dominated by Great Britain, but that Britain lost out by mismanagement to the United States in what he calls the second or "Henry Ford" stage of the industrial revolution. He argues that Japan or possibly Europe may overtake the United States in the next period if we continue to keep our economic focus on production of goods instead of on production and distribution of information.

The most comprehensive analysis of the role of information in the U.S. Economy was reported by economist Fritz Machlup in his 1962 book, *The Production and Distribution of Knowledge in the United States.* Working with 1958 data, he concluded (p. 362) that the production and distribution of knowledge accounted for 29% of the gross national product (GNP) in 1958. He included education, research and development, media of communication, information machines, and information services in that total.

[...]

According to Machlup's calculations, this total consists of 80 per cent Final Product (including both investment and consumption) and 20 per cent Intermediate Product (current costs of production of other products). He does not attempt a formal breakdown between investment and consumption, but argues that the expenditures on information as investment exceed the consumption expenditures on information.

Using two different estimation techniques, one based on incomes of workers in knowledge-producing occupations and one based on growth rates of knowledge industries, Machlup concludes that the information segment of the economy is growing at the rate of approximately 10 per cent per year, a rate approximately double that of the economy as a whole.

Data taken from the National Income Tables presented in the *Survey of Current Business,* U.S. Department of Commerce, provide a crude indication of the growth of information-related goods and services in the private sector of the U.S. economy between 1950 and 1970. Table 1 presents the general summary by major sector of the economy. Despite increases in dollar amounts, the percentage of national income derived from manufacturing declined from 31% in 1950 to 27% in 1970. Agriculture declined from 7% of national income in 1950 to 3% in 1970.

Parker presents a series of data tables showing changes in the components of four sectors of the economy that can be interpreted as indicative of information products and services. Even excluding government expenditures, which are largely information related and which increased from 10 per cent of the national income in 1950 to 16 per cent in 1970, the trend shows an increase of information expenditures from 9.7 per cent in 1950 to 12.2 per cent in 1960 and 14.3 per cent in 1970. This factor-of-1.5 increase in the percentage of national income and factor-of-five increase in unadjusted current dollars (from $23 billion to $115 billion) indicates how strongly information-related activities dominated the economic growth of the past two decades. These trends are likely to be accelerated in the 1970's, pushed along by the expansion of cable television, video cassettes, computer hardware and services,

TABLE 1

Components of National Income (current $)[1]

	1950 (Bil.)	1950 (%NI)	1960 (Bil.)	1960 (%NI)	1970 (Bil.)	1970 (%NI)
NI (billions)	$239.17		$414.50		$798.63	
Agriculture	17.38	7%	17.30	4%	25.55	3%
Mining	4.99	2%	5.51	1%	7.67	1%
Construction	12.40	5%	21.79	5%	42.79	5%
Manufacturing	74.50	31%	121.02	29%	216.23	27%
Wholesale/retail	42.78	18%	67.70	17%	121.19	15%
Finance	20.53	9%	42.59	10%	89.96	11%
Transportation	13.20	6%	17.91	4%	29.69	4%
Communication/ utilities	7.16	3%	16.81	4%	31.41	4%
Services	22.31	10%	49.06	12%	102.66	13%
Government	23.37	10%	52.53	13%	126.80	16%
Rest of the world	.55	0%	2.29	1%	4.62	1%

[1] Source: Survey of Current Business, U.S. Department of Commerce.

communication satellites, and other advances in communication technology.

Another way of studying economic trends is by reviewing labor statistics. Data from the Bureau of the Census on the composition of the work force in 1950, 1960, and 1970 can be compared with projections of the U.S. Bureau of Labor Statistics for 1980. Professional, technical, and managerial occupations (and white-collar occupations generally) are more concerned with processing of information than with direct handling of products. The trends and projections for this are shown in Table 2.

TABLE 2

Percent of Labor Force in Specified Occupations

	1950	1960	1970	1980 (projected)
Professional, Technical & Managerial	17.0%	21.8%	24.7%	27.8%
All White Collar Workers	37.5%	43.1%	48.3%	50.1%

The Bureau of Labor Statistics has also estimated the percentage increase to be expected in various occupations between 1968 and 1980. The increase in total number of white collar jobs is estimated at 50 per cent. The number of jobs in which an advanced degree is now considered necessary for professional advancement is estimated to increase by 38 per cent. In the job categories most concerned with new information technology (i.e., computer programmers, systems analysts, computer operating personnel, etc.), a whopping 145 per cent increase between 1968 and 1980 is predicted. These are the occupations that make technical training, rather than advanced-degree training occupations, the fastest growing in the work force. (Blue collar and unskilled labor jobs are the ones in relative decline.)

These statistics provide a simple economic description of the dominant trend in the society: a shift from an industrial society to an information society. Such facts underlie the various qualitative discussions of what is described as the "post-industrial society" or the "knowledge society."

Why is this trend taking place? The most likely explanation is that expenditures on information may constitute the most promising investment in improved economic productivity. Increasing productivity means producing more goods and services for the same amount of human labor. This usually means working "smarter" rather than working harder (i.e., applying new information to the production processes).

Information, rather than labor or capital, is becoming the key factor in production. Peter Drucker argues, "knowledge, during the last few decades, has become the central capital, the cost center, and the crucial resource of the economy". Servan-Schreiber, in *American Challenge*, outlines what he sees as the United States's threat of economic domination of western Europe as essentially an information advantage. The recent Japanese "white paper" titled *The Plan for Information Society – a National Goal Toward Year 2000*, recommends a major, centrally planned development of what they call "the information society." They propose a five-year investment of 1,000 billion yen (3.2 billion dollars). Their argument is that with such a national investment they can sustain an annual rate of growth of GNP in excess of 10 per cent per year, contrasted with a 7 per cent growth rate if they follow a U.S. style laissez-faire policy of information investment.

A recent U.S. congressional report titled *American Productivity: Key to Economic Strength and National Survival* also recommends investment in an expanded communication infrastructure as a key to U.S. economic growth. That report says (p.6), "Although a skeleton national computer net exists ... expansion of such a net into a fullfledged computer utility, like other regulated utilities, could bring vast productivity gains at low costs in services where information exchange is basic."

One form of investment in information is expenditure on education. Denison attributes 23 per cent of the economic growth of the United States from 1929 to 1957 to increases in the level of education. He further attributes 20 per cent of the economic growth in the same period to advances in knowledge. When these figures are translated into economic growth per person employed, by dividing by the size of the labor force, 42 per cent of the economic growth is attributed to improvements in education and 36 per cent to advances in knowledge. Phelps argues that optional investment in education becomes increasingly larger the more technologically progressive is the society. These arguments for social investment in education also apply to the public and other libraries that constitute a significant component of the total educational services available in society.

Will (and should) this trend of increasing expenditures on information continue? Given the goal of continued economic growth, then investment in information should continue to increase in total amount and in the percentage of the total economy. The argument for increases in the percentage of gross national product, not just increasing dollar amounts to maintain a stable percentage, follows from Phelps's argument for increased need for information as the society becomes technologically more complex.

But most of our economic productivity gains have been accomplished in the agricultural and manufacturing sectors of the economy. With manufacturing now providing only about a quarter of our national income and agriculture only about 3 per cent, we may be reaching a point of diminishing returns. With total information expenditures by public and private sectors of the economy now exceeding expenditures on manufacturing and agriculture, the traditional growth arguments may not apply. Further increases in present types of information expenditures may not lead to as large an overall growth if the sectors in which the growth is expected to appear constitute an already small and further declining percentage of the total economic activity of the society.

Because of this diminishing-returns problem in an increasingly information-dominated society, two questions need to be analyzed more carefully than has been necessary in the past. First, more attention should be paid to the efficiency of expenditures on information. We are no longer an information-poor society in which most additional information services could be justified as leading to economic growth. We are becoming an information-rich society in which we need to consider more carefully which information services are most likely to lead to economic growth. The relative efficiency of different forms of expenditure in information services should be carefully analyzed. It would be ridiculous to be inefficient in the expenditures in a larger sector of the economy (information) while claiming the goal is improved efficiency in a smaller sector of the economy (manufacturing).

The second major question raised by these economic trends concerns the need for investment focused on productivity gains in the information sector of the economy itself. If information is, as we have argued, an already large and fast-growing component of the economy, then the best chance for productivity gains in the economy as a whole may be investment in research and development and in education that can lead to productivity gains within the information sector of the economy itself. Just as education and research and development led to dramatic economic gains in the agricultural and manufacturing sectors of the economy (i.e., in the use of matter and energy), so the appropriate road to further national economic growth may be through similar investment in research and education leading to more efficient and effective use of information itself.

Efficient information investment

What form of information expenditure is more effective? The first of our two major questions is more difficult. Better answers to this question would be one of the desired outcomes from increased expenditure on information about information, as suggested in our second major question. But some indications are available from previous research on this question. For a recent general collection of articles on this subject, see *Economics of Information and Knowledge* (Lamberton).

Perhaps the most important question for both economists and policy makers is the extent to which efficient allocation of resources is obtained through the private sector of the economy (i.e., through free-market mechanisms). In a classic paper, Nobel prize-winning economist Kenneth Arrow argues that a free-enterprise economy can be expected to underinvest in information production for three reasons. One is the "indivisibility" of information; a second is inappropriability (the discoverer or inventor cannot obtain all the economic benefits from his discovery); and the third is uncertainty. His arguments are made in the context of technical invention, but they can be broadened to include investment in information production generally.

With respect to distribution of information, even that most consistent advocate of free markets, conservative economist Milton Friedman, in his section on education in *Capitalism and Freedom*, argues that a free market leads to underinvestment in education. He argues for both equity financing by the federal Government (with payments returned through income tax) and subsidy in the form of fees good at any approved institution as ways of compensating for market imperfections.

The same reasons that education is primarily a governmental activity apply to libraries and other information institutions as well. Nothing in the current economic trends is changing that need for continued governmental involvement in financing information services. Which level of government can most effectively supply those services is a question outside the scope of this chapter.

Who needs the information services most? Becker's data indicate that the least-well-educated (or least-well-informed) can most benefit (in the direct sense of increased dollar income) from increased education or information. There is a diminishing marginal return as higher levels of education are attained. Unfortunately for library policy and financing, the more educated and more informed are more likely to recognize the value of additional information. People with less education are less likely to have savings over and above current consumption needs and are less likely to be able to borrow money to invest in their own education, or even to know the value of such education.

Well-educated middle-class taxpayers may be the ones most likely to provide the political support for public library financing, even though economic-efficiency arguments would state that service policies aimed at those more economically and educationally deprived would be more socially beneficial than services desired by middle-class patrons. The better-educated segments of the population may also be better able to afford information available through the private sector of the economy (e.g., by buying paperback and hard-cover books). Information for the information-poor is more likely to be of benefit for both the individual and the society as a whole. The information-poor are also economically poor and hence less able to obtain information from the private sector of the economy.

Some may argue against major investment in information resources for the economically and culturally deprived segments of American society. They maintain that such people do not want or would not use the information and would use such facilities only or primarily for entertainment.

This argument is flawed because it ignores the considerable elasticity of the demand for information. When entertainment is inexpensive (provided "free" by television because of advertising) and education and information have high social and economic costs, it is not surprising that people in economically deprived sectors of the society spend little of their limited energy and resources on investment in information to improve their future. They have enough trouble meeting the basic needs of the present. Conversely, if the society makes education and information resources as freely and easily available to all as television entertainment is now available, the use of those resources could increase considerably. To compare present use of "free" television (excluding the cost of the television set) and free libraries would be unfair on two counts. One is that much more money is spent (by advertisers) to subsidize television than is spent (by government) to subsidize libraries, and television has more incentive to work hard at attracting audiences.

The second reason that the comparison is unfair is that libraries are far from free when the time and effort required for access by the user is included in the calculation. Becker's economic theory of the allocation of time provides a basis for a fairer comparison. This is not to argue that large numbers of people will suddenly switch from comic books to the *New York Times* or from the want ads to computer-aided instruction in the economic theory of international monetary transactions. The information will have to be perceived as relevant as well as inexpensive before it will be widely used. Understanding one's rights under the welfare or social security system, obtaining information from "Dear Abby" or "Dr. Hippocrates," or learning basic job skills, are examples of information likely to be in greater demand. The only way one can maintain the argument that information would not be used is to assume that information will continue to be as expensive to obtain as it is now or to assume that the information will be irrelevant to the interests of those not now benefiting from existing information resources.

Another flaw in the pessimistic argument is that it depends on a stereotype that denies the variability across individuals. It falsely assumes that all will react the same way. It is likely that some will not make any use of improved information and education resources and that others will make only minimal use. But still others are likely to make use of the opportunities presented. The economic argument for investment in human resources dictates that no people should be denied the additional opportunities that extended information systems can present. The implications of this economic argument coincide with those of the moral argument that we should not deprive some individuals of the benefits of improved information systems merely because others may not use them, or may not use them fully. Suppose a similar debate were held soon after Gutenberg's development of the printing press. A strong case could be made against widespread diffusion on the grounds that the illiterate public would not use it anyway. The force of the argument would have been stronger then than now, but it is equally flawed, as history has shown.

Providing public library service for the least-well-educated and least-well-informed segments of the population may imply a shift in the kinds of media and services provided by libraries. More emphasis on audio and video media, with a consequent relative decline of emphasis on print, may be called for. Shifts in geographic location and in library staffing may be required to make services more accessible both physically and socially. Different kinds of reference service and educational programs may be required to meet the needs of less-well-educated segments of the population.

Information research and development

The second major question raised by the trends in the economy concerns the importance of investment in activities intended to make the information sector of the economy itself more productive. Productivity gains and economies of scale in manufacturing have led to lowered unit costs of most manufactured goods. Information services, such as education and library services, generally have not benefited from similar cost-reducing changes in their means of production and distribution. Pake discussed the crisis in higher education in the United States and pointed to the productivity problem as a key issue. He cited costs per unit of instruction that increased by a factor of four during the decade of the 1960's in many universities. He said a major challenge to educational institutions is to find a way to use technology to increase productivity. The same comments can be applied equally well to education at all levels, including public education services of libraries.

This is not just a problem for libraries and other educational institutions. It is a national economic problem of considerable magnitude. The rate of economic growth of the United States is lower than that of Japan and other countries. This points to a relative decline of the United States as a world power unless ways are found to improve the rate of economic growth. Investment in an improved information infrastructure may be essential for a smooth transition from a manufacturing society to a knowledge society. Development of information utilities that permit economical on-demand access to information services from most homes and offices may be the most promising road to national economic growth, especially as the pressures increase to stabilize our consumption of matter

and energy. [...] the advent of two-way cable television, communication satellites, and computer information systems can permit the construction of a national information system that could stimulate rapid economic growth. Although investment costs would be high, both in the research and development needed and in the capital costs of installing such a system, the gains are likely to be large. The extension of information services now provided at libraries (on-demand information and education) to most homes and offices at costs the society can afford may be the most significant factor influencing the U.S. economy in the last quarter of the present century.
[...]

Policy implications

The general economic and technological trends influencing the United States are certain to have a major impact on libraries and other institutions currently concerned with information production and distribution. Because so many of the expected shifts in the economy and technology are changes in the amount and the costs of information processing, traditional institutions concerned with information transfer should structure their policy planning within the general context of those trends. Otherwise, the policies that will have the most effect on libraries will be made elsewhere and libraries will be reduced to the role of passive observers of trends they are not effectively influencing or controlling.

The general areas of policy consideration can be subsumed under four general headings. First is national science policy. The second is education policy. The third is technology policy. The fourth is library policy. The first three are more likely to determine the role and function of libraries in society than is the fourth.

National science policy

Recent changes in U.S. science policy have emphasized a shift away from a policy of supporting basic science as the primary goal of science policy. The new emphasis is on the development of science and technology that will help solve national problems or lead to increased economic growth. President Nixon's announcement in January 1973 of the abolition of the Office of Science and Technology and the President's Scientific Advisory Committee signalled a formalization of the shift. The transfer of responsibility for Federal science policy to the National Science Foundation (NSF), with NSF reporting to the chairman of the Council on Economic Policy, is a clear indication of the new context for science policy. The reorganization included establishment of a Federal Science Council, reporting to the director of NSF, to coordinate all government agency research and development programs, and the establishment of the Science and Engineering Council to provide NSF with advice from academic and industrial scientists and engineers. The rapid growth within

NSF of the program on Research Applied to National Needs (RANN) is another example of the general science policy.

This general shift in policy calls for a corresponding shift in policies for scientific and technical information. Much of earlier science information policy was concerned with information support for science and technology. Now a larger and even more challenging mission has been added: to develop information policies that will facilitate the application of scientific and technical information to the solution of national problems and to facilitate the national economic growth. In other words, in addition to maintenance of present information systems in support of science, information systems need to be developed to improve the flow of information between science and the rest of society. That information system needs a two-way flow, so that society can maximally benefit from the "production of knowledge" and so that the scientific community better understands the problems for which they are asked to seek solutions.

A recent report by the Organization for Economic Co-operation and Development (OECD) contains a series of recommendations for national policy on scientific and technical information. That report, *Information for a Changing Society – Some Policy Considerations*, makes it clear that such policies must be formulated in the context of a more general approach to the problems of information and society. The transmittal letter from study group chairman Pierre Pigainol to the OECD Secretary-General focuses on the central question of economic growth. Pigainol says, "What are these hard facts? Obviously the central one is economic growth ... The problem of scientific and technical information seems to be closely linked with economic growth ..."

If the economic arguments concerning the growing importance of information for economic productivity are heeded, an expanded program of research and development on the role of information in society is called for. Investment in the science and technology of information processing may provide as much of a spur to economic development as investment in information services for science and technology. One sub-area of research on information deserving special attention at this time is the economics of information. There are too few economists working in this area, and many promising avenues to explore. The goal of a cohesive program of research on the economics of information should be better understanding of the relationship between investment in information services and economic growth. [...]

Education policy

The need for economical open access to education by every member of U.S. society throughout his or her lifetime is now being articulated because it is technically possible to meet that need in the next two decades. The type of on-demand access to self-study materials that has characterized libraries since their earliest days can be extended on a broader scale throughout society through electronic media. As the technology develops and understanding of its potential spreads throughout the society, the need for such material will be expressed more stridently.

The development of facilities and institutions for alternate education or "life-long learning" will create special demands on libraries. Open universities will require expanded service offerings by local libraries. Planning for open university services or experiments should include plans for increasing library resources to support such activities. One form of open schooling that could be experimented with is the development of learning modules (in print, audio, or video media) that could be made available through public libraries for self-study programs.

Development of a national examining university to give academic credit or recognition to learning through self-study programs would provide a major stimulus for and recognition of such programs. Developing an examining university that does not, itself, give courses may be the most economically productive option in the range of alternate education proposals.

A recent Carnegie Commission report and reports by the HEW Task Force examine the relationship between education and information technology. The *JCET News,* newsletter of the Joint Council on Educational Telecommunications, Inc., edited in Washington, D.C. by Frank Norwood, provides current news concerning educational telecommunication questions.

Information technology policy

Libraries have for the past several years been discussing prospects for library networks. Unfortunately, specialized networks linking libraries together are usually expensive and seldom can be shown to produce benefits that justify the costs. If general-purpose, broadband electronic networks were available for communication of data and text throughout society, the social and economic benefits could be enormous. Library networking would constitute only a tiny fraction of the use of general networks, with business and other government services providing the bulk of the traffic. The resulting economies of scale would mean that library networks could be provided at an affordable incremental cost. The main justification for development of such networks would be for the general economic growth and improved social services that could result.

A recent report titled *Information Technology: Initiatives for Today – Decisions that Cannot Wait* presents a set of policy recommendations, many requiring Federal action, prepared under the direction of a prestigious organization of influential business executives, the Senior Executives Council of the National Industrial Conference Board. Their concern is with information policy questions that cover a much broader area than libraries. With respect to libraries, they say (p.5), "What happens, for instance, when existing public information systems, such as libraries, are made obsolete by a network of automatic storage and retrieval systems?" The recent report is Part Two of a major study of information technology. Part One, published in January 1972, titled *Information Technology: Some*

Critical Implications for Decision Makers, consisted of nine background papers discussing probable development in the information environment during the 1970's and 1980's.

The primary motivation for their policy recommendations is the improvement of U.S. economic productivity. One is left with the impression that building a broadband digital communications infrastructure for the society and developing the kinds of information services such an infrastructure would permit is the major opportunity for U.S. economic growth in the next decades.

The Conference Board report argues that implementation of a coordinated national information policy is likely to produce significantly higher rates of economic growth than would otherwise obtain. This opportunity (especially when viewed in the competitive context of Japanese and European economic growth rates) may constitute the most significant information policy question for the United States. The opportunity is for investment in information technology for economic growth. It appears possible to develop a technological information infrastructure that significantly lowers the unit costs of storing, retrieving, processing, and transmitting information.

Development of a technology-based information infrastructure of broadband information networks could permit significant lowering of unit costs for education and public information as well as private sector information services. The opportunity is analogous to the railroad-building era of a century ago. Railroads were to a mass-production society what information networks will be to a knowledge society. The difference is that railroads require much energy and are used to transport matter. An information infrastructure, since it is concerned with the transmission of ideas, can be consistent with a balanced ecology while permitting improvements in the quality of life. But, as was the case in the development of transcontinental railroads, government participation will be needed in policy formation and in the provision of leverage funds to influence the direction of developments.

The policies needed for development of broadband information networks will require many novel features. Much of the investment will be made by the private sector of the economy. Federal funding may be appropriate for research and development activities and to pay for costs of public-sector services (such as library interconnection and other education services) that use the networks. Many policies will be concerned with structuring of private markets such that sufficient incentives are available for orderly economic development. Attention will have to be given to questions of technical standards and to regulatory policy (e.g., by the Federal Communications Commission). Federal policy may be more effective, however, if the influence on private investment is more frequently accomplished by positive economic leverage than by negative regulatory sanctions. Voluntary standards may be sufficient without the need for mandatory standards if government purchasing power is used to provide incentives to meet the standards.

[...]

Library policy

Several conclusions regarding library services could be drawn from the economic and technological trends discussed above.

Expanded audio and video services should be provided in response to the general shift toward greater use of such media that will occur during the rest of this decade. Greater emphasis on information for the information-poor will be necessary to counterbalance partially the likely widening of the gap between the information-rich and the information-poor that will result from increased commercial development and exploitation of information technology (including pay television). Switching centers and referral services should be developed so that libraries can come closer to meeting the widening diversity of information needs, even though it may be uneconomical to provide a full range of service in each local library. National service to local libraries (e.g., on-line computerized searches of the Library of Congress MARC files) could be provided (analogous to MEDLINE) to make national bibliographic information readily accessible throughout the country.

Consideration should be given to improving the citizen's access to public information about government services and government decision-making at all levels. Minutes and supporting documents of all local government boards and committees could be made more accessible through local libraries for example. Within a few years, a national network of Federal Government information could be made available to local libraries via computer time-sharing and information retrieval techniques, just as medical references are made available to the medical libraries by the National Library of Medicine's MEDLINE system. Computerized congessional information systems now being developed could be made nationally accessible by the Library of Congress.

Conclusion

This chapter has taken a broad perspective on the role of information in society, approaching the problem primarily from an economics perspective. Evidence was presented indicating that information activities constituted an increasing percentage of the national income. Expanded investment in information appears to be a promising path to continued economic growth without requiring major increases in consumption of matter and energy. The trends in four currently changing facets of communication technology were examined. These are communication satellites, video recording, computer services, and cable television. Implications for national science policy, education policy, information technology policy and library policy were discussed. Government expenditure on information services and on information research and development were examined as potential "leverage" to provide incentives for the private sector of the economy to develop an expanded, general-purpose broadband communications infrastructure permitting improved information services in both the private and public sectors of the economy.

The post-industrial society: a conceptual schema

Daniel Bell

The concept "post-industrial society" emphasizes the centrality of theoretical knowledge as the axis around which new technology, economic growth and the stratification of society will be organized. Empirically one can try to show that this axial principle is becoming more and more predominant in advanced industrial societies.

This is not to suggest a principle of convergence. The idea of convergence is based on the premise that there is *one* overriding institution that can define a society. Thus, years ago, Pitirim Sorokin and C. Wright Mills thought that the Soviet Union and the United States were coming to be "alike" because both were becoming centralized, bureaucratic societies mobilized around the single purpose of war. In a different way, Jan Tinbergen and his associates have argued that, because of the nature of economic rationality, both communist and capitalist countries are veering toward a common model of modified planning (direct or indirect) tied to the use of markets. And Marion Levy and, to some extent, Wilbert Moore, have argued that in their central features all industrial societies were becoming "alike" because of the common requirements of factory production, the relation of education to occupation, and the character of technical knowledge. But few societies, as distinct historic and political entities, can be defined completely around a single institution as Marx believed, i.e. that one could define a system as capitalist and that *all other relations*, cultural, religious, political, derived from that foundational base. Societies differ in the way they relate their political systems to social structure and culture. And within the social structure (as in the political order and culture) there are different axes around which institutions are built. And even when different societies may be compared in their economy or social structure along one axis, they may differ markedly along another. Thus, along the axis of property, different societies could be defined as capitalist, but along the axis of, say, political consensus, these societies would differ as democratic and authoritarian.

In a different sense, one can say that the sequences of feudalism, capitalism and socialism, and pre-industrial and post-industrial society both proceed from Marx. Marx defined a mode of production as including both the social relations and the "forces" (i.e., techniques). He called the present mode of production capitalist, but if we restrict the term capitalist to social relations and industrial to techniques, then we can see, analytically, how the different sequences unfold. In this sense there can be socialist post-industrial societies as there could be capitalist, just as both the Soviet Union and the United States, though separated along the axis of property, are both industrial societies.

Further, one has to make a distinction between convergence and internationalization. One may have an internationalization of style, in painting or music or architecture, so that a "modern" artist may paint the same way in France, England, Japan, and Mexico. And there may be an internationalization of scientific knowledge and technological processes, but *societies,* as specific historical entities, represent distinct *institutionalized* combinations which are difficult to match directly to each other. On different dimensions (technology, architecture) they may resemble each other, or draw from a common fund of knowledge or style, yet on other dimensions (values, political systems, traditions) – and the ways these become formulated, say, in educational systems – they may differ. If there is a meaning to the idea of convergence it is that societies resemble each other somewhat along the same dimensions, or they may confront a *similar core of problems.* But this in no way guarantees *a common or like response.* The response will be relative to the different political and cultural organization of the specific society.

The idea of a post-industrial society, like that of an industrial society, or capitalism, has meaning only as a conceptual scheme. It identifies a new axial principle of social organization and defines a common core of problems which societies that become more and more post-industrial have to confront. The idea of a post-industrial society does not rest, as Jean Floud seems to think, on the concept of a social system. I do not believe that societies are organic or so integrated as to be analyzable as a single system. In fact, my major theoretical preoccupation today is the disjunction, in Western society, between the culture and the social structure, the one becoming increasingly anti-institutional and antinomian, the other oriented to functional rationality and meritocracy. The concept of post-industrialism is an effort to identify a change in the *social structure.* But there is no necessary correlate, as I have argued repeatedly, between the changes in that realm as against changes in the other two analytical dimensions of a society, that of politics and culture.

I do believe that the major orders, or realms, of societies can best be studied by trying to identify *axial* institutions or principles, which are the major lines around which other institutions are draped, and which pose the major problems of solution for the society. In capitalist society the axial institution has been private property and in the post-industrial society it is the centrality of theoretical knowledge. In Western culture in the last one hundred years the axial thread has been "modernism" with its onslaught on tradition and established institutions. In the Western political systems the axial problem is the relation between the desire for popular participation and bureaucracy.

One of the difficulties for social analysis is the overlapping and conflicting nature of these principles vis-à-vis one another, and within a single system itself. Thus, in the stratification system, which sociologists take as basic for any society, the historic base of power has been property, and the means of access through inheritance. Yet while property remains as an important base, technical skill becomes another, sometimes rival, base with education the means of access to the attainment of technical skill. But equally, a political office becomes a base for power and privilege (particularly for groups such as ethnic groups that find barriers in the other two modes), and political mobilization or cooptation becomes the means of access to political office. It is the contradictory nature of the three modes that makes identification of consistent social groups and political interests so difficult. Culture has replaced technology as a source of change in the society, and the tensions between the adversary culture and the eroded Protestant ethic have created a remarkable contradiction in the value system of American society.

In what way does a post-industrial society differ from previous society? In an historical sense, Professor Tominaga is right: the post-industrial society is a continuation of trends unfolding out of industrial society, and many of the developments were foreseen long ago. Both St. Simon and Marx, for example, were obsessed with the crucial role of engineers (in the one case) and science (in the other) in transforming society, though neither of them had, or could have, any sense of the change in the fundamental relation of science to economic and technological development, the fact, for instance, that most of the major industries of the nineteenth and early twentieth centuries – steel, telegraph, telephone, electricity, auto, aviation – were developed largely by talented tinkerers who worked independently of the fundamental work in science; while the first modern industry is chemistry, since one has to have a theoretical *a priori* knowledge of the properties of the macromolecules one is manipulating in order to create new products.

Yet for *analytical* purposes, one *can* divide societies into pre-industrial, industrial, and post-industrial and see them in contrast along many different dimensions. Thus, I have tried to show in the accompanying schematic table how, along different dimensions, the three kinds of societies differ radically. (See Table 1–1, General Schema of Social Change.) These are, of course, ideal-type constructions, yet the purpose of such constructs is to illuminate essential differences. Thus, the "design" of pre-industrial society is a "game against nature": its resources are drawn from extractive industries and it is subject to the laws of diminishing returns and low productivity; the "design" of industrial society is a "game against fabricated nature" which is centered on man-machine relationships and uses energy to transform the natural environment into a technical environment; the "design" of a post-industrial society is a "game between persons" in which an "intellectual technology," based on information, rises alongside of machine technology. Because of these different designs, there are vast differences in the character of economic sector distribution and the slopes of occupations. The methodology of each society differs and, most important, there are distinctly different axial principles around

TABLE 1–1

General Schema of Social Change

	Pre-industrial	Industrial	Post-industrial
Regions:	Asia Africa Latin America	Western Europe Soviet Union Japan	United States
Economic sector:	Primary Extractive: Agriculture Mining Fishing Timber	Secondary Goods producing: Manufacturing Processing	Tertiary Transportation Utilities Quaternary Trade Finance Insurance Real Estate Quinary Health Education Research Government Recreation
Occupational slope:	Farmer Miner Fisherman Unskilled worker	Semi-skilled worker Engineer	Professional and technical Scientists
Technology:	Raw materials	Energy	Information
Design:	Game against nature	Game against fabricated nature	Game between persons
Methodology:	Common sense experience	Empiricism Experimentation	Abstract theory: models, simulation, decision theory, systems analysis
Time perspective:	Orientation to the past Ad hoc responses	Ad hoc adaptiveness Projections	Future orientation Forecasting
Axial principle:	Traditionalism: Land/resource limitation	Economic growth: State or private control of investment decisions	Centrality of and codification of theoretical knowledge

which the institutional and organizational features of the society are draped.

Because of all this, there are crucial differences in the kinds of structural problems that engage each kind of society. In industrial society the chief economic problem has been the problem of capital: how to institutionalize a process of creating sufficient savings and the conversion of these moneys into investment areas; and this has been done through the equity market, investment banks, self-financing, and state taxation. The locus of social relations has been the enterprise or firm and the major social problem that of industrial conflict between employer and worker. To the extent that the investment process has been routinized and the "class conflicts" encapsulated so that the issue of class strife no longer acts to polarize a country around a single issue, those older problems of an industrial society have been muted if not "solved."

In the post-industrial society, the chief problem is the organization of science, and the primary institution the university or the research institute where such work is carried out. In the nineteenth and early twentieth centuries the strength of nations was their *industrial* capacity, the chief index of which was steel production. The strength of Germany before the First World War was measured by the fact that she had overtaken Great Britain in steel production. After the Second World War, the *scientific* capacity of a country has become a determinant of its potential and power, and research and development (R & D) has replaced steel as a comparative measure of the strength of the powers. For this reason the nature and kinds of state support for science, the politicization of science, the sociological problems of the organization of work by science teams all become central policy issues in a post-industrial society. (See Table 1–2, Structure and Problems of the Post-Industrial Society.)

A major social change brings a major reaction. The student revolts of the late 1960s were, in part, a reflection of the new power of an adversary

TABLE 1–2

Structure and Problems of the Post-Industrial Society

Axial principle:	The centrality of and codification of theoretical knowledge
Primary institutions:	University Academy institutes Research corporations
Economic ground:	Science-based industries
Primary resource:	Human capital
Political problem:	Science policy Education policy
Structural problem:	Balance of private and public sectors
Stratification: Base – Access –	Skill Education
Theoretical issue:	Cohesiveness of "new class"
Sociological reactions:	The resistance to bureaucratization The adversary culture

culture reacting against the growth of a science-based society. But in greater measure the student revolt was a reaction to the "organizational harnesses" a post-industrial society inevitably drops on intellectual work, and this was signified by the increasing pressures on young people, at an earlier and earlier age, to choose a good college, the pressure to choose a main subject, and the anxiety about graduate school and a career.

Politically, the problem of a post-industrial society, as François Bourricaud rightly points out, is the growth of a non-market welfare economics and the lack of adequate mechanisms to decide the allocation of public goods. For technical and conceptual reasons one cannot measure the value of such goods in market terms; because such goods are distributed to all, there is a disincentive on the part of a citizenry to support such expenditures. Most important, the nature of non-market political decisions invites direct conflict: as against the market which disperses responsibility, in politics the decision points are open and visible, and the consequences of political decisions, as to who would lose and who would gain, are clear. Because there is such a focal point, conflict flares more readily when such decisions are to be reached.

In the broadest sense, the most besetting dilemma confronting all modern society is bureaucratization, or the "rule of rules". Historically, bureaucratization was in part an advance of freedom. Against the arbitrary and capricious power, say, of a foreman, the adoption of impersonal rules was a guarantee of rights. But when an entire world becomes impersonal, and bureaucratic organizations are run by mechanical rules (and often for the benefit and convenience of the bureaucratic staff), then inevitably the principle has swung too far.

All of these changes take place in the context of a society that is multiplying its constituencies (especially the scientific and technical claimants), that mixes technocratic and political decisions, and that sees the rise of a new class which may or may not be struggling to establish a corporate cohesiveness as a new ruling class in society. It is all these issues that frame the problems of a post-industrial society.

The concept of a post-industrial society is not a picture of a complete social order; it is an attempt to describe and explain an axial change in the social structure (defined as the economy, the technology and the stratification system) of the society. But such a change implies no specific determinism between a "base" and a "superstructure"; on the contrary, the initiative in organizing a society these days comes largely from the political system. Just as various industrial societies – the United States, Great Britain, Nazi Germany, the Soviet Union, post-World War II Japan – have distinctively different political and cultural features, so it is likely that the various societies that are entering a post-industrial phase will have different political and cultural configurations. The essential division in modern society today is not between those who own the means of production and an undifferentiated "proletariat" but the bureaucratic and authority relations between those who have powers of decision and those who have not, in all kinds of organizations, political, economic, and social. It becomes the task of the political system to manage these relations in response to the various pressures for distributive shares and social justice.

What the concept of a post-industrial society suggests is that there is a common core of problems, hinging largely on the relation of science to public policy, which will have to be solved by these societies; but these can be solved in different ways and for different purposes. The sociologist seeks those "ordering devices" that allow one to see the way social change takes place in a society. The concept of the post-industrial society is one such "ordering device" to make more intelligible the complex changes in Western social structure.

What Tawney defined is *professionalism,* and if this vision is correct, the heart of the post-industrial society is a class that is primarily a professional class. As with any status group, the boundaries of definition are fluid and often indistinct, yet certain core elements are obvious. A profession is a learn*ed* (i.e. scholarly) activity, and thus involves formal training, but within a broad intellectual context. To be within the profession means to be certified, formally or informally, by one's peers or by some established body within the profession. And a profession embodies a norm of social responsiveness. This does not mean that professionals are more charitable or high-minded than their fellows, but that expectations about their conduct derive from an ethic of service which, as a norm, is prior to an ethic of self-interest. For all these reasons, the idea of a profession implies an idea of competence and authority, technical and moral, and that the professional will assume an hieratic place in the society.

One might say that business is called to account by its customers through the market, whereas a professional is called to account by his peers through the professional group. Property is associated with wealth, which can be passed on through legal title directly; a profession is defined by skill which can be passed on only indirectly through cultural advantage that the children of professionals may gain.

[...] If one turns to the societal structure of the post-industrial society [...], two conclusions are evident. First, the major class of the emerging new society is primarily a professional class, based on knowledge rather than property. But second, the control system of the society is lodged not in a successor-occupational class but in the political order, and the question of who manages the political order is an open one. (See "Schema: The Societal Structure of the Post-Industrial Society" [*below*].

In terms of status (esteem and recognition, and possibly income), the knowledge class may be the highest class in the new society, but in the nature of that structure there is no intrinsic reason for this class, on the basis of some coherent or corporate identity, to become a new economic interest class, or a new political class which would bid for power. The reasons for this are evident from an inspection of the Schema.

The professional class as I define it is made up of four estates: the scientific, the technological, the administrative, and the cultural. While the estates, as a whole, are bound by a common ethos, there is no intrinsic interest that binds one to the other, except for a common defense of the idea of learning; in fact there are large disjunctions between them.

Schema: The Societal Structure
of the Post-Industrial Society (U.S. Model)

I. *Statuses: Axis of Stratification — Based on Knowledge* (Horizontal Structures)
  A. The professional class: the four estates
    1. Scientific
    2. Technological (applied skills: engineering, economics, (medicine)
    3. Administrative
    4. Cultural (artistic and religious)
  B. Technicians and semi-professional
  C. Clerical and sales
  D. Craftsmen and semi-skilled (blue-collar)

II. *Situses: Locations of Occupational Activities* (Vertical Structures)
  A. Economic enterprises and business firms
  B. Government (bureaucratic: judicial and administrative)
  C. Universities and research institutions
  D. Social complexes (hospitals, social-service centers, etc.)
  E. The military

III. *Control System: The Political Order*
  A. The directorate
    1. Office of the President
    2. Legislative leaders
    3. Bureaucratic chiefs
    4. Military chiefs
  B. The polities: constituencies and claimants
    1. Parties
    2. Elites (scientific, academic, business, military)
    3. Mobilized groups
      a) Functional groups (business, professional, labor)
      b) Ethnic groups
      c) Special-focus groups
        (1) Functional (mayors of cities, poor, etc.)
        (2) Expressive (youth, women, homosexual, etc.)

The scientific estate is concerned with the pursuit of basic knowledge and seeks, legitimately, to defend the conditions of such pursuit, untrammeled by political or extraneous influence. The technologists, whether engineers, economists, or physicians, base their work on a codified body of knowledge, but in the application of that knowledge to social or economic purposes they are constrained by the policies of the different situses they are obedient to. The administrative estate is concerned with the management of organizations and is bound by the self-interest of the organization itself (its perpetuation and aggrandizement) as well as the implementation of social purposes, and may come into conflict with one or another of the estates. The cultural estate – artistic and religious – is involved with the expressive symbolism (plastic or ideational) of forms and meanings, but to the extent that it is more intensively concerned with meanings, it may find itself increasingly hostile to the technological and administrative estates. As I noted in the introduction, the axial principle of modern culture, in its concern with the self, is antinomian and anti-institutional, and thus hostile to the functional rationality which tends to dominate the application of knowledge by the technological and administrative estates. Thus in the post-industrial society one finds increasingly a disjunction between social structure and culture which inevitably affects the cohesiveness if not the corporate consciousness of the four estates.

While the classes may be represented, horizontally, by *statuses* (headed by the four estates), the society is organized, vertically, by *situses*, which are the actual loci of occupational activities and interests. I use this unfamiliar sociological word *situses* to emphasize the fact that in day-to-day activities the actual play and conflict of interests exist between the organizations to which men belong, rather than the more diffuse status identity and ethos. In a capitalist society, the property owner or businessman, as a class, is located exclusively in the business firm or corporation, so that status and situs are joined. In the post-industrial society, however, the four estates are distributed among many different situses. Scientists can work for economic enterprises, government, universities, social complexes, or the military (though the bulk of the "pure" scientists are to be found in the university). And the same distributions hold for the technologists and the managers. Because of this "cross-cutting," the likelihood of a pure "estate" consciousness for political purposes tends to diminish.

Finally, if the major historical turn in the last quarter-century has been the subordination of the economic function to societal goals, the political order necessarily becomes the control system of the society. But who runs it, and for whose (or what) ends? In one respect, what the change may mean is that traditional social conflicts have simply shifted from one arena to another, so that what the traditional classes fought out in the economic realm, where men sought comparative advantage in place, privilege and domination, is now transferred to the political realm, and as that arena widens, the special foci and ethnic groups (the poor and the blacks) now seek to gain through politics the privileges and advantages they could not obtain in the economic order. This is what has been taking place in recent years, and it will continue. The second, and structurally more pervasive, shift is that in the post-industrial society the *situses* rather than the *statuses* would be the major political-interest units in the society. To some extent this is evident in the familiar phenomenon of pressure groups. But in the post-industrial society it is more likely that the *situses* will achieve greater corporate cohesiveness vis-à-vis one another and become the major claimants for public support and the major constituencies in the determination of public policy. And yet the very forces which have re-emphasized the primacy of the political order in a technical world make it imperative to define some coherent goals for the society as a whole and, in the process, to articulate a public philosophy which is more than the sum of what particular situses or social groups may want. In the efforts to forge some such coherence one may find the seeds of the cohesiveness of the professional class in the post-industrial society.

[...]

It is quite striking that the one society in modern history which consciously began with a principle of almost complete equality (including almost no wage differentials) – the Soviet Union – gradually abandoned that policy, not because it was restoring capitalism but because it found that differential wages and privileges served as incentives and were also a more rational "rationing" of time. (If a manager's time is worth more than that of an unskilled worker, since he has to make more decisions, should he be expected to wait in line for a crowded tram or be given a car of his own to get to work?) Even those societies which have had relatively small differentials in income and incentives in the post-World War II years, such as Israel and Yugoslavia, have gradually widened these differences in order to stimulate productivity. And one of the chief pieces of advice which sympathetic economists have given to Fidel Castro to restore his stumbling economy (which has been largely organized on the basis of moral exhortation and the donation of extra labor time) is to make greater use of material incentives and wage differentials. In the United States, the major period when social programs could be most easily financed was from 1960 to 1965, when the increase in the rate of economic growth, not the redistribution of income, provided a fiscal surplus for such programs.

The United States today is not a meritocracy; but this does not discredit the principle. The idea of equality of opportunity is a just one, and the problem is to realize it fairly. The focus, then, has to be on the barriers to such equality. The redress of discrimination by representation introduces arbitrary, particularistic criteria which can only be destructive of universalism, the historic principle, won under great difficulty, of treating each person as a person in his own right.

The difficult and thorny question, in the end, is not just priority – who should be helped first – but the degree of disparity among persons. How much difference should there be in income between the head of a corporation and a common laborer, between a professor at the top of the scale and an instructor? The differences in pay in a business firm are on the order of 30:1, in a hospital of 10:1, and a university of 5:1. What is the rationale for these differences? What is fair? Traditionally, the market was the arbiter of differential reward, based on scarcity or on demand. But as economic decisions become politicized, and the market replaced by social decisions, what is the principle of fair reward and fair differences? Clearly this will be one of the most vexing questions in a post-industrial society.

A striking fact of Western society over the past two hundred years has been the steady decrease in the disparity among persons – not by distribution policies and judgments about fairness, but by technology, which has cheapened the cost of products and made more things available to more people. The irony, of course, is that as disparities have decreased, as democracy has become more tangible, the expectations of equality have increased even faster, and people make more invidious comparisons ("people may suffer less but their sensibility is exacerbated"), a phenomenon now commonly known as the "Tocqueville effect." The revolution of rising expectations is also the revolution in rising *ressentiment*.

The real social problem, however, may be not the abstract question of "fairness" but the social character of *ressentiment*, and the conditions which give rise to it. The fascinating sociological puzzle is why in democratic society, as inequality decreases, *ressentiment* increases. That, too, is part of the ambiguous legacy of democracy.

A just meritocracy

The difficulty with much of this discussion is that inequality has been considered as a unitary circumstance, and a single principle the measure of its redress, whereas in sociological fact there are different kinds of inequality. The problem is not *either/or* but what *kinds* of inequality lead to what *kinds* of social and moral differences. There are, we know, different kinds of inequality – differences in income and wealth, in status, power, opportunity (occupational or social), education, services, and the like. There is not one scale but many, and the inequalities in one scale are not coupled completely with inequality in every other.

We must insist on a basic social equality in that each person is to be given respect and not humiliated on the basis of color, or sexual proclivities, or other personal attributes. This is the basis of the civil rights legislation outlawing modes of public humiliation such as Jim Crow laws, and setting forth the principle of complete equal access to all public places. This principle also makes sexual conduct a purely private matter between consenting adults.

We should reduce invidious distinctions in work, whereby some persons are paid by the piece or the hour and others receive a salary by the month or year, or a system whereby some persons receive a fluctuating wage on the basis of hours or weeks worked and others have a steady, calculable income.

We should assert that each person is entitled to a basic set of services and income which provides him with adequate medical care, housing, and the like. These are matters of security and dignity which must necessarily be the prior concerns of a civilized society.

But one need not impose a rigid, ideological egalitarianism in all matters, if it results in conflict with other social objectives and even becomes self-defeating. Thus, on the question of wage or salary differentials, there may be good market reasons for insisting that the wages of a physician and dentist be greater than those of a nurse or dental technician, for if each cost the patient roughly the same (if one could for the same price have the services of a better qualified person) no one would want to use a nurse or dental technician, even in small matters. The price system, in this case, is a mechanism for the efficient rationing of time. If as a result of differential wages the income spread between the occupations became exceedingly high, one could then use the tax laws to reduce the differences.

But the point is that these questions of inequality have little to do with the issue of meritocracy – if we define the meritocracy as those who have an *earned* status or have achieved positions of rational authority by competence. Sociologists have made a distinction between power and authority. Power is the ability to command, which is backed up, either implicitly or explicitly, by force. That is why power is the defining principle of politics. Authority is a competence based upon skill, learning, talent, artistry or some similar attribute. Inevitably it leads to distinctions between those who are superior and those who are not. A meritocracy is made up of those who have earned their authority. An unjust meritocracy is one which makes these distinctions invidious and demeans those below.

Contemporary populism, in its desire for wholesale egalitarianism, insists in the end on complete levelling. It is not *for fairness,* but *against elitism;* its impulse is not justice but *ressentiment.* The populists are for power ("to the people") but against authority – the authority represented in the superior competence of individuals. Since they lack authority, they want power. In the populist sociology, for example, the authority of doctors should be subject to the decisions of a community council, and that of professors to the entire collegiate body (which in the extreme versions include the janitors).

But there cannot be complete democratization in the entire range of human activities. It makes no sense, in the arts, to insist on a democracy of judgment. Which painting, which piece of music, which novel or poem is better than another cannot be subject to popular vote – unless one assumes, as was to some extent evident in the "sensibility of the sixties", that all art is reducible to experience and each person's experience is as meaningful to him as anyone else's. In science and scholarship, achievement is measured and ranked on the basis of accomplishment – be it discovery, synthesis, acuity of criticism, comprehensive paradigms, statements of new relationships, and the like. And these are forms of intellectual authority.

All of this underscores a confusion between a technocracy and a meritocracy. Because the technocratic mode reduces social arrangements to the criterion of technological efficiency, it relies principally on credentials as a means of selecting individuals for place in the society. But credentials are mechanical at worst, or specify minimum achievement at best; they are an entry device into the system. Meritocracy, in the context of my usage, is an emphasis on individual achievement and earned status as confirmed by one's peers.

Rawls has said that the most fundamental good of all is self-respect. But the English sociologist W. G. Runciman has made a useful distinction between respect and praise. While all men are entitled to respect, they are not all entitled to praise. The meritocracy, in the best meaning of that word, is made up of those worthy of praise. They are the men who are the best in their fields, as judged by their fellows.

And just as some individuals are worthy of praise, so are certain institutions – e.g. those engaged in the cultivation of achievement, the institutions of science and scholarship, culture and learning. The university is dedicated to the authority of scholarship and learning and to the transmission of knowledge from those who are competent to those who are capable. There is no reason why a university cannot be a meritocracy, without impairing the esteem of other institutions. There is every reason why a university has to be a meritocracy if the resources of the society – for research, for scholarship, for learning – are to be spent for "mutual advantage," and if a degree of culture is to prevail.

And there is no reason why the principle of meritocracy should not obtain in business and government as well. One wants entrepreneurs and innovators who can expand the amount of productive wealth for society. One wants men in political office who can govern well. The quality of life in any society is determined, in considerable measure, by the quality of leadership. A society that does not have its best men at the head of its leading institutions is a sociological and moral absurdity.

Nor is this in contradiction with the principle of fairness. One can acknowledge, as I would, the priority of the disadvantaged (with all its difficulty of definition) as an axiom of social policy, without diminishing the opportunity for the best to rise to the top through work and effort. The principles of merit, achievement, and universalism are, it seems to me, the necessary foundations for a productive – and cultivated – society. What is important is that the society, to the fullest extent possible, be a genuinely open one.

The question of justice arises when those at the top can convert their authority positions into large, discrepant material and social advantages over others. The sociological problem, then, is how far this convertibility is possible. In every society, there are three fundamental realms of hierarchy – wealth, power, and status. In bourgeois society, wealth could buy power and deference. In aristocratic society, status could command power and wealth (through marriage). In military and estate societies, power could command wealth and status. Today it is uncertain whether the exact relations between the three any longer hold: Income and wealth (even when combined with corporate power) rarely command prestige (who knows the names, or can recognize the faces, of the heads of Standard Oil, American Telephone, or General Motors?); political office does not make a man wealthy; high status (and professors rank among the highest in prestige rankings) does not provide wealth or power. Nor does the existence of a meritocracy preclude the use of other routes – particularly politics – to high position and power in the society.

But even within these realms, the differences can be tempered; and the politics of contemporary society makes this even more likely in the future. Wealth allows a few to enjoy what many cannot have; but this difference can – and will – be mitigated by a social minimum. Power (not authority) allows some men to exercise domination over others; but in the polity at large, and in most institutions, such unilateral power is increasingly checked. The most difficult of all disparities is the one of status, for what is at stake is the desire to be different and to *enjoy* the disparity. With his usual acuteness into the passions of the human heart, Rousseau observed: "[It is] the universal desire for reputation, honors and preferences, which devours us all, trains and compares talents and strengths . . . stimulates and multiplies passions; and making all men competitors, rivals or rather enemies, how many reverses, successes and catastrophes of all kinds it causes. . ."

Yet, if vanity – or ego – can never be erased, one can still observe the equality of respect due to all, and the differential degree of praise bestowed upon some. As Runciman puts it, "a society in which all inequalities of prestige or esteem were inequalities of praise would to this extent be just." It is in this sense that we can acknowledge differences of achievement between individuals. It is to that extent that a well-tempered meritocracy can be a society if not of equals, then of the just.

[...]

Conclusion

The post-industrial society, as I have emphasized several times, is primarily a change in the character of social structure – in a dimension, not the total configuration of society. It is an "ideal type", a construct, put together by the social analyst, of diverse changes in the society which, when assembled, becomes more or less coherent when contrasted with other conceptual constructs. In descriptive terms, there are three components: in the economic sector, it is a shift from manufacturing to services; in technology, it is the centrality of the new science-based industries; in sociological terms, it is the rise of new technical elites and the advent of a new principle of stratification. From this terrain, one can step back and say more generally that the post-industrial society means the rise of new axial structures and axial principles: a changeover from a goods-producing society to an information or knowledge society; and, in the modes of knowledge, a change in the axis of abstraction from empiricism or trial-and-error tinkering to theory and the codification of theoretical knowledge for directing innovation and the formulation of policy. Any massive set of social changes poses new problems of social management, and in this Chapter I have attempted to suggest the agenda of issues which emerges with the onset of post-industrial society: the new hierarchies of technical elites and the bureaucratization of science; meritocracy and equality; the antinomian thrust of an adversary culture, the communal society and the difficulty of consensus. These run the gamut from ethos and values to politics and social organization.

Yet, in an intangible way, there may be more – a change in consciousness and cosmology, the dark tinge of which has always been present at the edges of man's conception of himself and the world, which now moves to the phenomenological center.

In existentialist terminology, man is "thrown" into the world, confronting alien and hostile powers which he seeks to understand and master. The first confrontation was with nature, and for most of the thousands of years of man's existence, life has been a game against nature: to find shelter from the elements, to ride the waters and the wind, to wrest food and sustenance from the soil, the waters, and other creatures. The coding of much of man's behavior has been shaped by his adaptability to the vicissitudes of nature. In the nature of societal design, most of the world's societies still live in this game against nature.

Man as *homo faber* sought to make things, and in making things he dreamt of remaking nature. To be dependent on nature was to bend to its caprices and acknowledge its tyrannies and diminishing returns. To rework nature, to make fabricated things, was to enhance man's powers. The industrial revolution was, at bottom, an effort to substitute a technical order for the natural order, an engineering conception of function and rationality for the haphazard ecological distributions of resources and climates. In the industrial society, the cosmological vision was the game against fabricated nature.

The post-industrial society turns its back on both. In the salient experience of work, men live more and more outside nature, and less and less with machinery and things; they live with and encounter one another. The problem of group life, of course, is one of the oldest difficulties of human civilization, going back to the cave and the clan. But of necessity the context has changed. The oldest forms of group life were within the context of nature, and the overcoming of nature gave an external purpose to the lives of men. The group life that was hitched to things gave men a huge sense of power as they created mechanical artifacts to transform the world. But now these older contexts have been routinized, indeed have almost disappeared from men's view. In the daily round of work, men no longer confront nature, as either alien or beneficent, and fewer now handle artifacts and things. The post-industrial society is essentially a game between persons.

Will this changed experience create a change in consciousness and sensibility? For most of human history, *reality was nature,* and in poetry and imagination men sought to relate the self to the natural world. Then *reality became technics,* tools and things made by men yet given an independent existence outside himself, the reified world. Now *reality is primarily the social world* – neither nature nor things, only men – experienced through the reciprocal consciousness of self and other. Society itself becomes a web of consciousness, a form of imagination to be realized as a social construction. Inevitably, a post-industrial society gives rise to a new Utopianism, both engineering and psychedelic. Men can be remade or released, their behavior conditioned or their consciousness altered. The constraints of the past vanish with the end of nature and things.

But what does not vanish is the duplex nature of man himself – the murderous aggression, from primal impulse, to tear apart and destroy; and the search for order, in art and life, as the bending of will to harmonious shape. It is this ineradicable tension which defines the social world and which permits a view of Utopia that is perhaps more realistic than the here-and-now millennium on earth that modern man has sought.

Utopia has always been conceived as a design of harmony and perfection in the relations between men. In the wisdom of the ancients, Utopia was a fruitful impossibility, a conception of the desirable which men should always strive to attain but which, in the nature of things, could not be achieved. And yet, by its very idea, Utopia would serve as a standard of judgment on men, an ideal by which to measure the real. The modern *hubris* has sought to cross that gap and embody the ideal in the real; and in the effort the perspective of the ideal has become diminished and the idea of Utopia has become tarnished. Perhaps it would be wiser to return to the classic conception.

Men in their imagination will always seek to make society a work of art; that remains an ideal. Given the tasks that have to be solved, it is enough to engage in the sober construction of social reality.

The knowledge economy

Peter F. Drucker

The 'Knowledge Industries' which produce and distribute ideas and information rather than goods and services, accounted in 1955 for one-quarter of the U.S. Gross National Product. (This was already three times the proportion of the national product that the country had spent on the 'knowledge sector' in 1900.) Yet by 1965, ten years later, the knowledge sector was taking one-third of a much bigger national product. In the late 1970s it will account for one-half of the total national product. Every other dollar earned and spent in the American economy will be earned by producing and distributing ideas and information and will be spent on procuring ideas and information.

From an economy of goods, which America was as recently as World War II, we have changed into a knowledge economy.

The figures are impressive enough. Ninety per cent of all scientists and technologists who ever lived are alive and at work today. In the first 500 years since Gutenberg, from 1450 to 1950, some 30 million printed books were published in the world. In the last twenty-five years alone an equal number has appeared. Thirty years ago, on the eve of World War II, semi-skilled machine operators, the men on the assembly line, were the centre of the American work force. Today the centre is the knowledge worker, the man or woman who applies to productive work ideas, concepts, and information rather than manual skill or brawn. Our largest single occupation is teaching, that is, the systematic supply of knowledge and systematic training in applying it.

In 1900 the largest single group, indeed still the majority, of the American people, was rural and made a living on the farm. By 1940, the largest single group, by far, were industrial workers, especially semi-skilled (in fact, essentially unskilled) machine operators. By 1960, the largest single group were what the census called 'professional, managerial, and technical people', that is, knowledge workers. By 1975 or, at the latest by 1980, this group will embrace the majority of Americans at work in the civilian labour force.

Since the knowledge worker tends to be a good deal better paid than the manual worker, and also to have much greater job security, knowledge has already become the central cost of the American economy. The productivity of knowledge has already become the key to productivity, competitive strength, and economic achievement.

But the statistics, impressive though they are, do not reveal the important thing. What matters is that knowledge has become the central 'factor of production' in an advanced, developed economy.

Economists still tend to classify the 'knowledge industries' as 'services'. As such, they contrast them with the 'primary' industries – agriculture, mining, forestry, and fishing, which make available to man the products of nature – and with the 'secondary' industries – that is, manufacturing. But knowledge has actually become the 'primary' industry, the industry that supplies to the economy the essential and central resource of production. The economic history of the last hundred years in the advanced and developed countries could be called 'from agriculture to knowledge'. Where the farmer was the backbone of any economy a century or two ago – not only in numbers of people employed, but in importance and value of what he produced – knowledge is now the main cost, the main investment, and the main product of the advanced economy and the livelihood of the largest group in the population.

Increasingly knowledge is the key factor in a country's international economic strength. Increasingly we hear discussions of the 'brain drain', which pulls educated people from countries of relative knowledge backwardness towards countries of advanced knowledge status.

The best known example is the 'brain drain' from Britain to the United States – which is certainly closely connected with the fact that of the major industrial countries, Britain today has the lowest educational level for the great bulk of the population, with 80 per cent of her people leaving school at the age of 15. It is also the country in which access to productive livelihoods is still gained primarily through apprenticeship, that is, through experience rather than through knowledge.

But the 'brain drain' may be an even more serious problem for the underdeveloped countries whose educated people tend to go to work in the advanced industrial nations, especially in Britain and the United States (and to a lesser extent in France) rather than stay at home.

Similarly during the last few years there has been more and more talk of the 'technology gap' that is supposed to have opened up between the United States and Western Europe. There has been a good deal of fear, even by European friends of the United States, lest Europe become technologically subservient (a fear, for instance, expressed in the brilliant book of the French writer and journalist, Jean-Jacques Servan-Schreiber, *Le Défi Americain (The American Challenge)* which became a European best-seller in 1967.

In 1900 steel was the economic measurement – and the steel industry was based squarely on skill rather than on knowledge. If all the men of knowledge at work in a nation in 1900 had suddenly been taken away, the economy would hardly have noticed the difference. Knowledge was then ornament, economically speaking, rather than functional. Today it is the foundation and measurement of economic potential and economic power.

'Knowledge' rather than 'science' has become the foundation of the modern economy. To be sure, science and scientists have suddenly moved into the centre of the political, military, and economic stage.

Economists still tend to classify the 'knowledge industries' as 'services'. As such, they contrast them with the 'primary' industries – agriculture, mining, forestry, and fishing, which make available to man the products of nature – and with the 'secondary' industries – that is, manufacturing. But knowledge has actually become the 'primary' in dustry, the industry that supplies to the economy the essential and central resource of production. The economic history of the last hundred years in the advanced and developed countries could be called 'from agriculture to knowledge'. Where the farmer was the backbone of any economy a century or two ago – not only in numbers of people employed, but in importance and value of what he produced – knowledge is now the main cost, the main investment, and the main product of the advanced economy and the livelihood of the largest group in the population.

But so have practically all the other knowledge people. It is not just the chemists, the physicists, and the engineers who get fat on consulting assignments in universities – to the point where they may have a larger income from consulting outside the university than from teaching and research inside. Geographers and mathematicians, economists and linguists, psychologists, anthropologists, and marketing men all are busy consulting with government, with industry, with the foreign aid programme, and so on. Few areas of learning are not in demand by the organizations of our pluralist society. There is, I admit, little call for the consulting services of the Classics faculty. But there is more demand for the theologians than most people realize. Altogether it is the exceptional area of knowledge which is not today being brought into play in business and industry, in government and the military, in the hospital and in international relations.

This demand in return, reflects the basic fact that knowledge has become productive. The systematic and purposeful acquisition of information and its systematic application, rather than 'science' or 'technology', are emerging as the new foundation for work, productivity, and effort throughout the world.

This development has gone furthest in the United States. But in this area, as in so many others, the United States only started moving a few years earlier than the others. The trend runs the same way in every one of the advanced industrial countries. There is a close correlation between the ability of an economy to grow and compete in the world today and the rate of increase of its population at school beyond the age of fifteen – with the United States, Japan, Israel, and the Soviet Union at the top, and Great Britain at the bottom among the developed industrial nations. When the Europeans complain about the 'brain drain' and the 'technology gap', they are only asserting that their economies are not sufficiently 'knowledge-based' to perform, to grow, and to compete.

The demand for knowledge workers in the future seems insatiable. In addition to a million computer programmers, the information industry in the United States will need in the next fifteen years another half-million systems engineers, systems designers, and information specialists. We need, perhaps, two-million health care professionals nurses, dieticians, medical and X-ray technologists, social and psychiatric case workers, physiotherapists, and so on. These people are both highly trained, well above secondary school level, and highly skilled. They are fully the equivalent of the skilled machinist or the skilled carpenter with his years of apprenticeship. But their skill is founded on knowledge.

Maintenance of the air fleets of tomorrow – the jumbo jets and jet freighters which, between now and 1980, are likely to become the main carriers of people and freight – will require more workers than now keep all the railroads running. These men, while highly skilled, will greatly differ from the traditional railroad craftsman. First, they will be able to do the entire maintenance operation. They will not, like their predecessors, have been trained in a specific craft, whether sheet metal work or electronics, but in a specific function: keeping an aeroplane operating safely. Second, their skill will rest on theoretical knowledge and formal schooling rather than on apprenticeship in a craft. While they will work with their hands, they will apply knowledge rather than skill. Manuals, charts, and texts will be at least as important to them as the traditional hand tools are to the artisan.

These examples bring out some fundamentals of the knowledge economy.

(1) Knowledge work does not lead to a 'disappearance of work'. Eminent doctors tell us today that work is on its deathbed in the rich, industrially advanced countries, such as the United States, Western Europe, or Japan. The trends are actually running in the opposite direction. The typical 'worker' of the advanced economy, the knowledge worker, is working more and more, and there is a demand for more and more knowledge workers. The manual worker, the typical worker of yesterday, may have more leisure. He may go home at five in the evening. But the knowledge worker everywhere works increasingly longer hours: The young engineer, the accountant, the medical technologist, or the teacher, take work home with them when they leave the office. Knowledge work, like all productive work, creates its own demand. And the demand is apparently unlimited.

(2) Knowledge does not eliminate skills. On the contrary, knowledge is fast becoming the foundation for skill. We are using knowledge increasingly to enable people to acquire skills of a very advanced kind fast and successfully. Knowledge without skill is unproductive. Only when knowledge is used as a foundation for skill does it become productive. Then it enables us to acquire in less time and with less effort what it took years of apprenticeship to learn. It enables us to acquire new skills, i.e. computer programming, which could never be acquired through apprenticeship alone. Knowledge, that is, the systematic organization of information and concepts, is therefore making 'apprenticeship' obsolete. Knowledge substitutes systematic learning for exposure to experience.

We learned in World War II, with respect to manual crafts, that we could compress years of apprenticeship into weeks, or at most months, of organized and systematic learning. Once we had converted the experience of a craft to a systematic 'programme', people without great native talent or intelligence, that is, average people, became highly skilled craftsmen in little time, and enjoyed the learning experience. Britain after Dunkirk demonstrated this first, followed on a larger scale by the United States in 1942–44. We demonstrated this in metal work such as welding or riveting in shipbuilding, in engineering work of all kinds, and in construction. The Armed Forces used systems and programmes as the foundation for a great variety of skills – clerical, supervisory, and medical. With this experience the shift to knowledge work and knowledge industry actually began.

The man or woman who has once acquired skill on a knowledge foundation has learned to learn. He can acquire rapidly new and different skills. Unlike apprenticeship, which prepares for one specific craft and teaches the use of one specific set of tools for one specific purpose, a knowledge foundation enables people to unlearn and to relearn. It enables them, in other words, to become 'technologists' who can put knowledge, skills, and tools to work, rather than 'craftsmen' who know how to do one specific task in one specific way.

Applying knowledge to work has a long history. The priests who thousands of years ago organized Egypt's agriculture – and indeed her entire political and social life – around the knowledge of the flood tides of the Nile, were 'knowledge workers' and applied knowledge to work.

But these were exceptions. Knowledge and work, until very recent times, were separate and rarely touched each other. Knowledge was desirable for its intrinsic beauty, and at best as conducive to wisdom (though the evidence for this old belief is not overwhelming). Work was based on experience – and this was as true, until a century or two ago, even of the work of the physician. As regards the lawyer, Mr Justice Holmes's dictum: 'The life of the law has not been logic, it has been experience', is as valid today as it was a century ago.

'Knowledge' as normally considered by the 'intellectual' is something very different from 'knowledge' in the context of a 'knowledge economy' or of 'knowledge work'. For the 'intellectual', knowledge is what is in a book. But as long as it is in the book, it is only 'information' if not mere 'data'. Only when a man applies the information to do something does it become knowledge. Knowledge, like electricity or money, is a form of energy that exists only when doing work. The emergence of the knowledge economy is not, in other words, part of 'intellectual history' as it is normally conceived. It is part of the 'history of technology', which recounts how man puts tools to work. When the intellectual says 'knowledge' he usually thinks of something new. But what matters in the 'knowledge economy' is whether knowledge, old or new, is applicable, e.g. Newtonian physics to the space programme. What is relevant is the imagination and skill of whoever applies it, rather than the sophistication or newness of the information.

The idea that knowledge, systematically acquired, could be applied systematically to work is no more than 200 to 250 years old. It first occurred to the long line of English tool makers and tool designers of the eighteenth and early nineteenth centuries, which culminated in the towering figure of Joseph Whitworth (1803-87). These men were not only great inventors without whose work modern industry and modern technology would never have come into being. They also set out to systematize what they knew about mechanical work and to build this knowledge into the tool. This not only led directly to 'engineering', that is, to the codification of the right way to do any particular task; it also changed work and work-force – the true beginning of the Industrial Revolution. The deliberate design of tools enabled the merely competent workman to turn out – predictably and effectively time after time – work of pre-established precision and uniformity. The famous 'go/no-go' gauges which Whitworth designed almost at the beginning of his career enabled every journeyman to know in a single operation what the work was like, what it should be like, and what was needed to bring it up to standard – and then to measure the results against the standards. They were the first 'programmes' ever designed, and to this day they are the most successful ones. They foreshadowed the 'knowledge economy' in that knowledge, and in the work of the English tool designers, became the foundation for skill and the means to acquire easily and fast what had, over the ages, been accessible only to near-genius. What formerly required a 'master' now required the 'skilled worker'.

The next step, and a completely different one, was taken in the United States with the Morrill Act of 1862, which established a land-grant college in every state of the Union. What was new was not the idea of research in agriculture and of developing new methods, new seeds, new breeds, and so on, but the idea of converting farming altogether from a practice into a discipline, and of making every farmer an agronomist and systematic technologist. This was a pure act of faith at the time. There was no precedent. Indeed there was little in human experience to make the idea sound plausible. For fifty years there were few results; but at about the time of World War I the land-grant college and the extension service began to have the hoped-for impact on farm work, farm output, and farm productivity. Since then the productivity of agriculture has been multiplied, and agriculture has changed its character. It used to be the way of life of the great majority of mankind, and a subsistence living. Now it has become a capital-intensive, mechanized, and 'scientific' industry. A small number of highly-trained men, equipped with expensive mechanical and managerial tools, produce something totally new in the world, farm surpluses. This is a greater change in culture, society, and economy than most of the technological innovations we marvel at.

The most important step towards the 'knowledge economy' was, however, Scientific Management – that is, the systematic application of analysis and study to manual work, first pioneered by Frederick W. Taylor (1856-1915) in the last decades of the nineteenth century. Taylor, for the first time in history, looked at work itself as deserving the attention of an educated man. Before, work had always been taken for granted, especially by the educated. If they ever thought of it, they knew that work had been ordained – by God or by nature – and that the only way to produce more was to work more and work harder. Taylor saw that this was false. The key to producing more was to 'work smarter'. The key to productivity was knowledge, not sweat.

Taylor did not start out (as most people believe who have never read his work) with ideas of efficiency or economy, let alone with the purpose of making a profit for the employer. He started out with a burning social concern and was deeply troubled by what he saw as a suicidal conflict between 'labour' and 'capital'. And his greatest impact has also been social. For Scientific Management (we would today probably call it 'systematic work study', and eliminate thereby a good many misunderstandings the term has caused) has proved to be the most effective idea of this century. It is the only basic American idea that has had worldwide acceptance and impact. Wherever it has been applied, it has raised the productivity and with it the earnings of the manual worker, and especially of the labourer, while greatly reducing his physical efforts and his hours of work. It has probably multiplied the labourer's productivity by a factor of one hundred.

But above all, Taylor's Scientific Management provided the way out of the impasse of the nineteenth century; opened a 'third way' between nineteenth-century capitalism and nineteenth-century socialism. It proved them both false. For both assumed as a law of nature that the economic pie was given and could not be increased, except by putting in more capital or working more and harder. Taylor showed that the economic pie could be enlarged rapidly by applying knowledge to work. This did not create the harmony Taylor had naïvely hoped for. But it replaced an irreconcilable conflict of principles which could only end by subjugating one to the other – either by a dictatorship of the 'capitalist' or by one of the 'proletariat' – with conflict over the division of the fruits of higher productivity. Fighting for 'more' can be bitter and protracted; but it can always be compromised.

It is not true that Taylor took the skill out of manual work, though this has often been asserted. For he applied Scientific Management only to work that had never been skilled, i.e. to work of the labourer. His most famous study, made in 1899, dealt with shovelling sand. And Schmidt, his shovelling labourer, was not skilled, had no pride in his non-existent craft, no control over his work, no fun doing it, and a bare subsistence to show for his ten hours of back-breaking, but largely unproductive, daily toil. Taylor's study thus enabled the unskilled man to be paid handsomely, almost at the level of the skilled man, and to be in high demand. The labourer suddenly became productive. Taylor, in other words, repealed the 'iron law of wages' under which the unskilled manual labourer had always lived. He did this by creating a skill that had never existed before, the skill of the 'industrial engineer'. It was the first skill to be firmly based on knowledge rather than on experience. The 'industrial engineer' of Scientific Management is the prototype of all modern 'knowledge workers' – and to this day, one of the most productive ones.

(3) But while knowledge eliminates neither work nor skill, its introduction does constitute a real revolution both in the productivity of work and in the life of the worker.

Perhaps its greatest impact lies in changing society from one of predetermined occupations into one of choices for the individual. It is now possible to make one's living, and a good living at that, doing almost anything one wants to do and plying almost any knowledge. This is something new under the sun.

Most of mankind through the ages has had no choice at all; son followed father. The Indian caste system only gave religious sanction to what was the norm for most people. Of course, there was always some mobility, upward and downward – even the caste system in India could not entirely prevent this. But these were the exceptions, the few lucky ones, the occasional highly gifted one, the victim of war and catastrophes, or the totally improvident who gambled or gave away whatever they inherited. And in a world in which most people eked out a bare subsistence on the land, being a peasant was for most of mankind the one and only occupation.

A century ago even the educated man could only make a living through knowledge in a few narrowly circumscribed 'professions': clergyman, physician, lawyer, and teacher, plus – the one newcomer – civil servant (engineers came in only at the end of the last century).

In 1930, a friend of mine, one of the ablest young men ever to graduate from a great Oxford college, was offered a junior fellowship in mathematics at All Souls, the most prestigious academic institution in Great Britain. Thereupon his intelligent and well-to-do middle-class family descended upon him in a body and persuaded him to turn down the offer and go into banking in the City instead.

They pointed out that the young man would have to earn his living and that there were only half a dozen jobs in England that paid a mathematician a living wage, primarily the few senior professorships at Oxford, Cambridge, and London. Obtaining one of these was as much a matter of being at the right place at the right time, when one of the incumbents died, as it was a matter of mathematical genius and merit. The man who failed to obtain one of these plums could at best hope to be a wretched schoolmaster drilling Euclid into numbskulls for a pittance. The point is not that the young man's family was timid. The point is that they were more nearly right than wrong. Opportunities for mathematicians were almost non-existent in England thirty-five or forty years ago as they were in most countries in the world.

Today, needless to say, opportunities for a mathematician are unlimited. He need not be a Newton or a Gauss to make a very good living doing what he enjoys doing. And the same holds true for almost every other branch of knowledge. In each there are more opportunities for productive, rewarding, and well-paid work than there are men and women available to fill them.

At the same time, access to education is becoming the birthright of people in advanced societies – and its absence the badge of 'class domination'. It is the absence of access to education which is now meant when people in the developing countries speak of 'colonial oppression' or 'neo-colonialism'. Education a hundred years ago was still a privilege. In about 1850 it first became an opportunity which the educational systems in the developed countries increasingly made available to the gifted and ambitious among the poor and 'under-privileged'.

Within the last twenty or thirty years, access to education has however become a right. Nowhere is it yet guaranteed in the American Constitution, but it is clearly as important today as any of the rights written into formal 'Bills of Rights'. When the U.S. Supreme Court outlawed 'separate but equal' education for the American Negro and ordered the integration of the schools in 1954, it clearly assumed that the right of access to education was as solemnly embedded in the American Constitution as any of the rights actually guaranteed therein.

Of course, there are still limits – and not only of ability, but of wealth, of the accident of location, and certainly of race, even in the richer countries. On the whole, we are rapidly moving from a society in which careers and occupations were determined, i.e. largely by the accident of birth, into one in which we take freedom of choice for granted.

The problem today is not the lack of choice, but the abundance thereof. There are so many choices, so many opportunities, so many directions that they bewilder and distract the young people. No sooner have they shown a passing interest in this or that area than they are encouraged to make it their life's career. There are simply not enough of anything – whether metallurgists or specialists in Oriental languages, whether statisticians or psychologists, whether systems engineers or botanists; all are in chronically short supply. The educated man or woman is overwhelmed. Sometimes I feel that my mathematician friend in Oxford in 1930 had it easier than his children today.

But even though a surfeit of choice can become burdensome, the human horizon has still been widened immeasurably

(4) Knowledge opportunities exist primarily in large organizations. While the shift to knowledge work has made possible large modern organizations, it is the emergence of these organizations – business enterprise, government agency, large university, research laboratory, hospital – that in turn has created the job opportunities for the knowledge worker.

The knowledge opportunities of yesterday were largely for independent professionals working on their own. Today's knowledge opportunities are largely for people working within an organization as members of a team, or by themselves. This is true even where the organization itself antedates the knowledge society, as do military and government service, university, and hospital. There is still the *Herr Professor* who looks upon the university as existing for him, and who works and teaches by himself as he pleases and at most with 'Assistants'. But this is not how most of the work in the university is done any more. It is done by teams, by 'interdisciplinary groups', and in organized research centres.

There is a charming picture of the British Home Office in Victorian times in Anthony Trollope's novel *John Caldigate.* The novel came out in 1879 – in a period of massive reforms in English local government that the Home Office pioneered. Yet the Home Office which Trollope – himself a civil servant and thoroughly familiar with the British government of his time – portrays, is essentially one man, totally anonymous, but very powerful, serving directly under the minister and assisted by a few clerks. In the bureaucratic Russia of the Tsars, things were apparently not very different judging by what Tolstoy tells us about the official life of that powerful bureaucrat, Anna Karenina's hapless husband. And even later, before World War I, the government agency was still essentially one or a few trained men, each working by himself.

No one has ever accused the Austro-Hungarian Monarchy of understaffing its government offices. Yet when my father entered Austrian government service in 1897, he joined a total of ten other 'gentlemen', each with a separate assignment, each reporting directly to an old and powerful chief of the bureau who, in turn, reported directly to the cabinet. There were simply not enough educated people available to provide a larger staff. Today's Austria, though only one-tenth of the size of the Austria of 1900, employs about 500 university-trained men in the same department, most of them working in large teams rather than on individual assignments.

The knowledge worker of today, in other words, is not the successor to the 'free professionals' of 1750 or 1900. He is the successor to the employee of yesterday, the manual worker, skilled or unskilled.

This is very substantial upgrading. But it also creates an unresolved conflict between the tradition of the knowledge worker and his position as an employee. Though the knowledge worker is not a 'labourer', and certainly not a 'proletarian', he is still an 'employee'. He is not a 'subordinate' in the sense that he can be told what to do; he is paid, on the contrary, for applying his knowledge, exercising his judgement, and taking responsible leadership. Yet he has a 'boss' – in fact, he needs to have a boss to be productive. And the boss is usually not a member of the same discipline but a 'manager' whose special competence is to plan, organize, integrate, and measure the work of knowledge people regardless of their discipline or area of specialization.

The knowledge worker is both the true 'capitalist' in the knowledge society and dependent on his job. Collectively the knowledge workers, the employed educated middle-class of today's society, own the means of production through pension funds, investment trusts, and so on. These funds are the true 'capitalists' of modern society with which no private individual, were he richer than Croesus, Rothschild, and Morgan combined, could compete. But individually the knowledge worker is dependent on his salary, on the pension benefits and health insurance that go with it, and altogether on having a job and getting paid. Individually he is an 'employee', even though there is no other 'employer' in our society.

But the knowledge worker sees himself as just another 'professional', no different from the lawyer, the teacher, the preacher, the doctor, or the government servant of yesterday. He has the same education. He has more income. He has probably greater opportunities as well. He may realize that he depends on the organization for access to income and opportunity, and that without the investment the organization has made – and a high investment at that – there would be no job for him. But he also realizes and rightly so, that the organization equally depends on him.

This hidden conflict between the knowledge worker's view of himself as a 'professional' and the social reality in which he is the upgraded and well-paid successor to the skilled worker of yesterday, underlies the disenchantment of so many highly educated young people with the jobs available to them. It explains why they protest so loudly against the 'stupidity' of business, of government, of the armed services, and of the universities. They expect to be 'intellectuals', and they find that they are just 'staff'. Since this holds true for organizations altogether and not just for this or that organization, there is nowhere to flee to. If they turn their backs on business and go to the university, they soon find out that this, too, is a 'machine'. If they turn from the university to government service, they find the same situation there.

What most of them fail to realize is that the choice is not between these irksome jobs – which are irksome precisely because they are 'jobs' – and an illusory freedom. The choice is between jobs that have opportunity and are paid well, and working in the potato or cotton patch, hoeing or weeding, sixteen hours a day for a bare subsistence. But it may be too much to expect young people to understand this. Certainly every single one of them would argue that this may be true for all the others, but he is the exception, the one who is qualified to be a true 'professional'.

This clash between the expectations in respect to knowledge jobs and their reality will become sharper and clearer with every passing year. It will make the management of knowledge workers increasingly crucial to the performance and achievement of the knowledge society. We will have to learn to manage the knowledge worker both for productivity and for satisfaction, both for achievement and for status. We will have to learn to give the knowledge worker a job big enough to challenge him, and to permit performance as a 'professional'.

But no matter how good a job we do in the management of the knowledge worker – and so far, we have barely begun to work in this area – his status, function, and position in modern society is certain to be a central problem, politically as well as socially. It is likely to be the social question of the developed countries for the twentieth and probably for the twenty-first century.

The emergence of knowledge work

How did this shift to knowledge society and knowledge economy come about?

The popular answer is: 'Because jobs have been getting more complex and more demanding.' But the right answer is: 'Because the working life-span of man has increased so greatly.' It is not the demand for labour but the supply that underlies this great transformation of society and economy. And this, in turn, also explains the social and economic problems posed by the emergence of knowledge. The arrival of the knowledge worker changed the nature of jobs. Because modern society has to employ people who expect and demand knowledge work, knowledge jobs have to be created. As a result, the character of work is being transformed.

The demands of most jobs have changed but little, at least until recent years.

There is nothing, for instance, in the work of an American salesgirl that explains why, thirty years ago, junior high school was considered adequate preparation for the work while today the applicant is expected to have finished high school and preferably a few years of college as well. Nor does today's salesgirl at the age of 18 or 20 produce any more sales than the 15-year-old salesgirl of 1935, or the 12-year-old salesgirl of 1910.

The typical foreman in a mass production plant in the United States in 1929 had gone to work at the age of 15, with six years of schooling. Ten years later, before the outbreak of World War II, the typical assembly-plant foreman already had a high school education.

Today it is increasingly taken for granted that he is a college graduate. Yet the job is exactly what it was forty years ago. If anything, it has become less demanding as more and more of it has either been reduced to a formula or been taken out of the foreman's hands and shifted to such specialists as personnel people, quality controllers, and production planners.

That the much-vaunted 'complexity of today's jobs' is a myth is also shown by comparing American practice with that of other advanced industrial countries. For while moving in the same direction, Europe and Japan are still well behind in the educational upgrading of the entire population. Jobs which call for a year or two of college education in the United States are being staffed in Germany, for instance, with men and women who have finished what corresponds roughly to the American junior high school (the German *Ober-Sekunda-Reife*). Yet there is no discernible difference in the demands made on the employee, or in his or her productivity.

The best example is found in Canada where there are two educational standards side by side. Toronto and industrial Ontario are on the same educational standard as the American Midwest on which they border. A few hundred miles to the East, Montreal and industrial Quebec are only now starting the 'educational explosion'. The jobs that, in Ontario, are being filled by high school graduates, preferably with one or two years of college, are staffed by the same employers – supermarket chains, for instance, or commercial banks or manufacturers – with junior high school graduates in Quebec. They are paid quite differently; yet there is not much difference in the jobs they do or in their productivity.

The direct cause of the up-grading of the jobs is, in other words, the up-grading of the educational level of the entrant into the labour force. The longer he or she stays at school, the more education will be required for entrance into a given job or occupation.

But the extension of the years at school is itself only an effect rather than a cause. It is the result of a long development which drastically changed working-life expectancy in the industrially advanced countries.

That life expectancy has gone up sharply in the advanced countries is common knowledge. But few people seem to have noticed that the years of working life have gone up even faster. Around the turn of the century, that is, in the lifetime of people still alive and working today, few members of the labour force could expect to work to full capacity past the age of 45 or 50. They were at least likely to be impaired physically by then.

Fifty years ago European middle-class mothers used to tell their children: 'A government job does not pay much. But you won't starve if you become an invalid before your time. And it does provide for your widow and the children you leave behind.' The earlier retirement programme, e.g. those of the military and civil services of Europe, expected the few survivors to retire after twenty-five to thirty years of working life, that is, before the age of 50. One of the oldest general retirement systems, that of the Japanese, still makes retirement at 55 compulsory for all but the people at the very top, with a lump-sum retirement payment equal to two years of salary. This was adequate as late as 1940. Few Japanese lived that long and those that reached the age of 55 were unlikely to live a great many more years beyond retirement. Today, when Japanese life expectancies and working-life spans are equal to those of the most advanced countries of the West, these retirement provisions have become absurd.

Working-life span half a century ago was already higher than it had been throughout history. Prior to 1850, there was no country where average life expectancy greatly exceeded 33 or 35 years, which limited average working life to 20 years. By 1914 these figures had risen, in the most advanced countries, to around 30 years of working-life expectancy, from starting work at the age of 15 to disablement or death aged 45 or so. Today working life extends to 65 for the great majority. If people were still expected to start work at about the age of 14, as was the case everywhere up to World War I, there would be a working-life span of 50 years for the great majority – two and a half times that of a century ago, and two thirds larger than it was only a generation ago. Even with our later age of entrance into the work force – 18 or 20 – the working-life span in the advanced countries today is twice that of a century ago and 50 per cent larger than it was at about the time of World War I.

This means that we have more than doubled the human capital of the economy in the last century. We have also more than doubled the earnings potential of the individual; for a lifetime's income of an individual is, of course, his annual income multiplied by the number of years he is capable of earning it. This is the biggest increase in what the economists call the 'stock of capital' ever recorded in economic history. It is also the greatest advance ever recorded in economic wealth and well-being – infinitely greater than anything shown in the economist's statistics of the supply of *goods*, which is what the economist usually means when he talks of the 'standard of living'.

Extending the working-life span has altered the ratio between the productive and the dependent members of the population. This may have been an even greater boost to productivity than the increase in the stock of capital. How much there is in a society does not depend solely on how much the individual produces; it depends also on how many people have to be supported by the productivity of one producer. In fact, productivity of an economy might perhaps best be expressed as productivity of the individual divided by the number of dependent people he has to support. This is the reason why the 'population explosion' is such a threat to the social and economic welfare of the developing nations. It increases the number of dependants, particularly small children below the productive age, so rapidly that even great increases in *per capita* production on the part of the producing population do not result in larger *per capita* income and higher standards of living.

The extension of working-life span in today's developed countries has, during the last hundred years, altered drastically the traditional balance between producers and dependants. Not only do people who survive remain self-supporting until a much later age. More important, they stay producers so much longer that even with a somewhat larger family, the number of dependants per producer has gone down sharply.

The farmer or artisan of 1800 or 1850 who had a thirty-year working life had to support four or five dependants all through this period. On an average, only two or three of his children reached manhood and became producers themselves. But in order to have two or three grown children, his wife (or wives) had to give birth to seven or eight – so that there were always young ones in the home who had to be supported. In addition, his own parents, if they survived, were likely to become dependent in their mid-forties, when he himself still carried a heavy burden of dependent children.

Today, with a working life of forty years or more, and with a child mortality reduced so sharply that babies can be expected to survive to manhood, the producer in the developed countries only has to support two or three dependants at any time, and only, as a rule, for the first half of his own productive life. In other words, extending working-life span amounts to quadrupling the economic potential of the same resource, the individual productive human being.

This tremendous enrichment made possible the general adoption of what had always been the rarest privilege of the exceedingly rich: the wife-and-mother was able to leave the labour force and stay at home. As a result, not having one's wife at work became the badge of respectability and affluence in the developed countries during the nineteenth century. Many English workers still consider it an insult to be asked whether their wives work – a self-respecting working man should keep his woman at home.

Meanwhile, of course, the wives of the truly affluent are again going back into the labour force. The steady rise of married women at work in the developed countries is one of the most significant and most universal economic phenomena of the last two decades. It is the result of the productivity that creates 'affluence' and thus makes it possible for even the working-class woman to have 'mechanical servants' who take physical labour and time out of housework. It is the result also of advances in fertility control and infant and child health care which make it possible for the family in the developed countries to 'plan' both the number and the spacing of their children. The mother is no longer tied to the home beyond the early years of marriage, and can go back to work when the children start going to school. This, a complete reversal of the earlier trend, again greatly alters the ratio between producers and dependants and, with it, the 'stock of capital' of the advanced economies.

Contrary to popular belief, the advances in medicine had little to do with the lengthening of working-life span. One main cause was undoubtedly the shift from farming as the occupation of the majority to where it now is, the occupation of a very small minority – fewer than 6 per cent of the United States population for instance. For farm work ages people fast, especially the farm work of sixty years ago, before electricity and modern machinery. Also in traditional agriculture disabling accidents occurred far more frequently than even in the most dangerous industrial pursuits. The farmer, and especially his wife, tended to be old and disabled by the time their children reached their teens.

But the shift from labourer to machine operator was hardly less important. Whether Irish or Chinese, the 'navvies' who built the American railroads with pick and shovel rarely lasted more than five years or so before they were disabled by accident, liquor, syphilis, or plain back-breaking toil. As late as 1900 seamstresses in the garment lofts of New York, the London East End, Paris, or Vienna, did not last much longer. Within ten years most of them had fallen victim to blindness or to TB, the twin scourges of their trade.

Scientific Agriculture and Scientific Management are, therefore, the true heroes. They are the two main factors in the extension of working-life span. The drop in infant mortality and the advances in public health might well have lengthened total life span, but working-life span would not have moved upwards much. For this, it was necessary firstly to make possible the feeding of large populations by a small number of truly productive farmers, and secondly, to raise productivity by eliminating the physically destructive tasks from industrial work.

The rise in working-life span, in turn, has led almost everywhere to a substantial extension of years spent in being educated.

It can be argued that this extension reflects a common faith in education as one of the highest human values. People are eager to spend a substantial part of new wealth such as a longer working-life span represents, on education rather than on getting an immediate money income.

It can also be argued that extending the years at school is an economic rationality in which people can indulge once they no longer have to worry where the next meal is coming from. By postponing earnings for a few years, they acquire higher earning power for the rest of their lives. There is no doubt that lifetime earnings rise disproportionately with education. If, in addition, working-life span also rises, the return on the investment in education, made by postponing the start of income by a few years, rises exponentially. In other words, extending the years of schooling is rational economic behaviour. It 'maximizes profits' far more effectively than anything the shrewdest businessman could ever have worked out.

There is one more explanation of this sharp increase in years of schooling. People cannot stand a working-life of fifty years. It is simply too long for them. A part of the gain in working-life span is therefore offset by postponing entrance into the labour force to a later age. School is not seen primarily as desirable in itself nor as a means to a better livelihood. It is seen very largely as a way of keeping the kids off the street while still keeping them out of the labour force for a few years.

There is plenty of evidence for this rather cynical view. Indeed there is evidence that even a working life of 45 years – that is, a working life that begins at the age of 20 – may be too long, especially for knowledge workers.

Whichever of these three explanations is preferred – and there are elements of all three in the situation – the results are the same. In the advanced countries, man has chosen to take a substantial part of the increased wealth made possible by Scientific Agriculture and Scientific Management in the form of longer years of schooling.

Extending the years of schooling forces us to create jobs which apply knowledge to work. The person who has sat on a school bench until 18 or 20 may not have learned anything. But he has acquired different expectations.

He (or she) expects in the first place a different kind of job, a job that is 'proper' for a high school or college graduate. This is firstly a job with higher pay. But also a job with greater opportunity. A job that gives a 'living' is no longer enough. It must offer a 'career'. But the job for the highly schooled is also a job in which one no longer works with one's hands but by applying one's mind. It is a knowledge job. Long years of schooling make a person unfit for anything but knowledge work. If he has learned anything at all, it is concept, system, idea; it is not experience. Finally, he is no longer available for apprenticeship; he is simply too old to start acquiring skill in the traditional way which takes five to eight years. The young man or woman who has sat in school until 18 or 20 cannot become a 'skilled worker', in the old style. Whatever he needs to learn must start from a foundation in knowledge. The 'education explosion' after World War II radically altered the supply of labour – first in the United States and increasingly in all the advanced industrial countries, including the Soviet Union. It made it impossible to continue the traditional job structure. In the United States in particular we were faced with the need to shift rapidly from manual jobs, skilled or unskilled, to knowledge jobs, skilled or unskilled, as the centre of employment. At the very least we had to upgrade the pay of a good many traditional jobs even though the jobs themselves did not change at all; the salesgirl is an example.

The basic problem in the American economy during the twenty years since World War II has never been the supply of manual jobs. This has been a concern only because of the special situation of the American Negro. For the white 90 per cent of the population, the problem has been to have adequate supply of knowledge jobs, or of jobs that paid on the knowledge scale. If we had not been able to justify the expectations of educated people for knowledge jobs, we would have today an unemployed and unemployable intellectual proletariat the like of which the world has never seen. It would present an infinitely more dangerous and more explosive problem than our racial ghettoes.

It is amazing that the American economy could satisfy the expectations of all these people with long years of schooling. Even more amazing is that American business, by and large, could raise productivity enough to maintain the competitive position of American industry in the world and the ability of the American economy to grow. For we did not have – and still do not have – much knowledge how to manage the knowledge worker and how to make him productive. In addition, in many cases, we had to pay much higher wages for people whose expectations were based on their long years of schooling, even though they did the same work with the same productivity as people with much less schooling – and with many fewer expectations – had done before.

As a result of the change in supply, we now have to create genuine knowledge jobs – whether the work itself demands it or not. For a true knowledge job is the only way to make highly schooled people productive.

By a circuitous and quite unplanned route, we are therefore, arriving at the point at which Taylor aimed seventy-five years ago: we are beginning to apply knowledge to work itself. That the knowledge worker came first and knowledge work second – that indeed knowledge work is still largely to come – is an historical accident. From now on, we can expect increasing emphasis on work based on knowledge, and especially on skills based on knowledge. The new jobs that open up in the economy will start out with theoretical and conceptual knowledge acquired systematically and on a 'course'. Old jobs will either be changed to knowledge jobs or will be replaced by knowledge jobs. We can expect, in other words, that we will develop a true knowledge economy, as we have already developed a knowledge work force. A good deal of this knowledge work, like computer programming, will be semi-skilled rather than highly skilled. But even then it will require a knowledge foundation acquired at school and college.

The social, political, and economic challenges in and from the work force in industrially advanced countries increasingly arise out of the dynamics of knowledge work and out of the needs of the knowledge worker.

Artificial intelligence

Joseph Weizenbaum

When Roger Schank expressed the hope that we will be able to build a program that can learn as a child does, he was echoing words spoken by H. A. Simon over ten years earlier:

> If GPS is a theory of how a machine can bootstrap itself into higher intelligence or how people learn language, then let it bootstrap itself, and let it learn language. This is an entirely appropriate obligation to impose. . . . Not just on behalf of myself, but on behalf of the entire group of people working in the field, I accept the obligation and hope that one of us will produce the requisite programs before too long.

Both Simon and Schank have thus given expression to the deepest and most grandiose fantasy that motivates work on artificial intelligence, which is nothing less than to build a machine on the model of man, a robot that is to have its childhood, to learn language as a child does, to gain its knowledge of the world by sensing the world through its own organs, and ultimately to contemplate the whole domain of human thought. (It is worth noting, though only by the way for now, that should this dream be realized, we will have a language-understanding machine but still no theory of language understanding as such, for observing a machine "learning as a child does" does not in itself constitute an understanding of the language-acquisition process.)

Whether or not this program can be realized depends on whether man really is merely a species of the genus "information-processing system" or whether he is more than that. I shall argue that an entirely too simplistic notion of intelligence has dominated both popular and scientific thought, and that this notion is, in part, responsible for permitting artificial intelligence's perverse grand fantasy to grow. I shall argue that an organism is defined, in large part, by the problems it faces. Man faces problems no machine could possibly be made to face. Man is not a machine. I shall argue that, although man most certainly processes information, he does not necessarily process it in the way computers do. Computers and men are not species of the same genus.

Few "scientific" concepts have so thoroughly muddled the thinking of both scientists and the general public as that of the "intelligence quotient" or "I.Q." The idea that intelligence can be quantitatively measured along a simple linear scale has caused untold harm to our society in general, and to education in particular. It has spawned, for example, the huge educational-testing movement in the United States, which strongly influences the courses of the academic careers of millions of students and thus the degrees of certification they may attain. It virtually determines what "success" people may achieve in later life because, in the United States at least, opportunities to "succeed" are, by and large, open only to those who have the proper credentials, that is, university degrees, professional diplomas, and so on.

When modern educators argue that intelligence tests measure a subject's ability to do well in school, they mean little more than that these tests "predict" a subject's ability to pass academic-type tests. This latter ability leads, of course, to certification and then to "success." Consequently, any correlation between the results of such tests and people's "success," as that term is understood in the society at large, must necessarily be an artifact of the testing procedure. The test itself has become a criterion for that with which it is to be correlated! "Psychologists should be ashamed of themselves for promoting a view of general intelligence that has engendered such a testing program."

My concern here is that the mythology that surrounds I.Q. testing has led to the widely accepted and profoundly misleading conviction that intelligence is somehow a permanent, unalterable, and culturally independent attribute of individuals (somewhat like, say, the color of their eyes), and moreover that it may even be genetically transmittable from generation to generation.

The trouble with I.Q. testing is not that it is entirely spurious, but that it is incomplete. It measures certain intellectual abilities that large, politically dominant segments of western European societies have elevated to the very stuff of human worth and hence to the *sine qua non* of success. It is incomplete in two ways: first, in that it fails to take into account that human creativity depends not only on intellect but also crucially on an interplay between intellect and other modalities of thought, such as intuition and wisdom; second, in that it characterizes intelligence as a linearly measurable phenomenon that exists independent of any frame of reference.

Einstein taught us that the idea of motion is meaningless in and of itself, that we can sensibly speak only of an object's motion relative to some frame of reference, not of any absolute motion of an object. When, in speaking informally, we say that a train moved, we mean that it moved relative to some fixed point on the earth. We need not emphasize this in ordinary conversation, because the earth (or our body) is to us a kind of "default" frame of reference that is implicitly assumed and understood in most informal conversation. But a physicist speaking as a physicist cannot be so sloppy. His equations of motion must contain terms specifying the coordinate system with respect to which the motion they describe takes place.

So it is with intelligence too. Intelligence is a meaningless concept in and of itself. It requires a frame of reference, a specification of a domain of thought and action, in order to make it meaningful. The reason this necessity does not strike us when we speak of intelligence in ordinary conversation is that the required frame of reference – that is, our own cultural and social setting with its characteristic domains of thought and action – is so much with us that we implicitly assume it to be understood. But our culture and our social milieu are in fact neither universal nor absolute. It therefore behoves us, whenever we use the term "intelligence" as scientists or educators, to make explicit the domain of thought and action which renders the term intelligible.

Our own daily lives abundantly demonstrate that intelligence manifests itself only relative to specific social and cultural contexts. The most unschooled mother who cannot compose a single grammatically correct paragraph in her native language – as, indeed, many academics cannot do in theirs – constantly makes highly refined and intelligent judgments about her family. Eminent scholars confess that they don't have the kind of intelligence required to do high-school algebra. The acknowledged genius is sometimes stupid in managing his private life. Computers perform prodigious "intellectual feats", such as beating champion checker players at their own game and solving huge systems of equations, but cannot change a baby's diaper. How are these intelligences to be compared to one another? They cannot be compared.

Yet forms of the idea that intelligence is measurable along an absolute scale, hence that intelligences are comparable, have deeply penetrated current thought. This idea is responsible, at least in part, for many sterile debates about whether it is possible "in principle" to build computers more intelligent than man. Even as moderate and reasonable a psychologist as George A. Miller occasionally slips up, as when he says, "I am very optimistic about the eventual outcome of the work on machine solution of intellectual problems. Within our lifetime machines may surpass us in general intelligence."

The identification of intelligence with I.Q. has severely distorted the primarily mathematical question of what computers can and cannot do into the nonsensical question of "how much" intelligence one can, again "in principle," give to a computer. And, of course, the reckless anthropomorphization of the computer now so common, especially among the artificial intelligentsia, couples easily to such simpleminded views of intelligence. This joining of an illicit metaphor to an ill-thought-out idea then breeds, and is perceived to legitimate, such perverse propositions as that, for example, a computer can be programmed to become an effective psychotherapist.

I had once hoped that it would be possible to prove that there is a limit, an upper bound, on the intelligence machines could achieve, just as Claude Shannon, the founder of modern information theory, proved that there is an upper bound on the amount of information a given information channel can transmit. Shannon proved that, for example, a specific telephone cable can carry at most a certain number of telephone conversations at any one time. However, before he could even sensibly formulate his now justly famous result, he had to have some way to quantify information. Else how could he speak of a channel's capacity to handle this "much" information but no "more"? Indeed, his design of an information measure itself constitutes an important contribution to modern science. (Given, of course, that he also founded a cogent theory within which his measure plays a decisive role.) It is now clear to me that, since we can speak of intelligence only in specific domains of thought and action, and since these domains are themselves not measurable, we can have no Shannon-like measure of intelligence and therefore no theorem of the kind I had hoped for. In plain words: we may express the wish, even the opinion, that there is a limit to the intelligence machines can attain, but we have no way of giving it precise meaning and certainly no way of proving it.

Does our inability to compute an upper bound on machine intelligence provide grounds either for the "optimistic" conclusion that "machines may surpass us in general intelligence" or for the very same "pessimistic" conclusion? Neither. We learn instead that any argument that calls for such a conclusion, or for its denial, is itself ill-framed and therefore sterile. [...]

The question I am trying to pursue here is, "What human objectives and purposes may not be appropriately delegated to computers?" We can design an automatic pilot, and delegate to it the task of keeping an airplane flying on a predetermined course. That seems an appropriate thing for machines to do. It is also technically feasible to build a computer system that will interview patients applying for help at a psychiatric out-patient clinic and produce their psychiatric profiles complete with charts, graphs, and natural-language commentary. The question is not whether such a thing *can* be done, but whether it is appropriate to delegate this hitherto human function to a machine.

The artificial intelligentsia argue, as we have seen, that there is no domain of human thought over which machines cannot range. They take for granted that machines can think the sorts of thoughts a psychiatrist thinks when engaged with his patient. They argue that efficiency and cost considerations dictate that machines ought to be delegated such responsibilities. As Professor John McCarthy once put it to me during a debate, "What do judges know that we cannot tell a computer?" His answer to the question – which is really just our question again, only in different form – is, of course, "Nothing." And it is, as he then argued, perfectly appropriate for artificial intelligence to strive to build machines for making judicial decisions.

The proposition that judges and psychiatrists know nothing that we cannot tell computers follows from the much more general proposition subscribed to by the artificial intelligentsia, namely, that there is nothing at all which humans know that cannot, at least in principle, be somehow made accessible to computers.

Not all computer scientists are still so naive as to believe, as they were once charged with believing, that knowledge consists of merely some organization of "facts". The various language-understanding and vision programs, for example, store some of their knowledge in the form of assertions, i.e., axioms and theorems, and other of it in the form of processes. Indeed, in the course of planning and executing some of their complex procedures, these programs compose subprograms, that is, generate new processes, that were not explicitly supplied by human programmers. Some existing computer systems, particularly the so-called hand-eye machines, gain knowledge by directly sensing their environments. Such machines thus come to know things not only by being told them explicitly, but also by discovering them while interacting with the world. Finally, it is possible to instruct computers in certain skills, for example, how to balance a broomstick on one of its ends, by showing them how to do these things even when the instructor is himself quite incapable of verbalizing how he does the trick. The fact, then, and it *is* a fact, that humans know things which they cannot communicate in the form of spoken or written language is not by itself sufficient to establish that there is some knowledge computers cannot acquire at all.

But lest my "admission" that computers have the power to acquire knowledge in many diverse ways be taken to mean more than I intend it to mean, let me make my position very clear.

First (and least important), the ability of even the most advanced of currently existing computer systems to acquire information by means other than what Schank called "being spoon-fed" is still extremely limited. The power of existing heuristic methods for extracting knowledge even from natural-language texts directly "spoonfed" to computers rests precariously on, in Winograd's words, "the tiniest bit of relevant knowledge". It is simply absurd to believe that any currently existing computer system can come to know in any way whatever what, say, a two-year-old child knows about children's blocks.

Second, it is not obvious that all human knowledge is encodable in "information structures", however complex. A human may know, for example, just what kind of emotional impact touching another person's hand will have both on the other person and on himself. The acquisition of that knowledge is certainly not a function of the brain alone; it cannot be simply a process in which an information structure from some source in the world is transmitted to some destination in the brain. The knowledge involved is in part kinesthetic; its acquisition involves having a hand, to say the very least. There are, in other words, some things humans know by virtue of having a human body. No organism that does not have a human body can know these things in the same way humans know them. Every symbolic representation of them must lose some information that is essential for some human purposes.

Third, and the hand-touching example will do here too, there are some things people come to know only as a consequence of having been treated as human beings by other human beings. I shall say more about this in a moment.

Fourth, and finally, even the kinds of knowledge that appear superficially to be communicable from one human being to another in language alone are in fact not altogether so communicable. Claude Shannon showed that, even in abstract information theory, the "information content" of a message is not a function of the message alone but depends crucially on the state of knowledge, on the expectations, of the receiver. The message "Am arriving on 7 o'clock plane, love, Bill" has a different information content for Bill's wife, who knew he was coming home, but not on precisely what airplane, than for a girl who wasn't expecting Bill at all and who is surprised by his declaration of love.

Human language in actual use is infinitely more problematical than those aspects of it that are amenable to treatment by information theory, of course. But even the example I have cited illustrates that language involves the histories of those using it, hence the history of society, indeed, of all humanity generally. And language in human use is not merely functional in the way that computer languages are functional. It does not identify things and words only with immediate goals to be achieved or with objects to be transformed. The human use of language manifests human memory. And that is a quite different thing than the store of the computer, which has been anthropomorphized into "memory." The former gives rise to hopes and fears, for example. It is hard to see what it could mean to say that a computer hopes.

These considerations touch not only on certain technical limitations of computers, but also on the central question of what it means to be a human being and what it means to be a computer.

I accept the idea that a modern computer system is sufficiently complex and autonomous to warrant our talking about it as an organism. Given that it can both sense and affect its environment, I even grant that it can, in an extremely limited sense, be "socialized," that is, modified by its experiences with its world. I grant also that a suitably constructed robot can be made to develop a sense of itself, that it can, for example, learn to distinguish between parts of itself and objects outside of itself, that it can be made to assign a higher priority to guarding its own parts against physical damage than to similarly guarding objects external to itself, and that it can form a model of itself which could, in some sense, be considered a kind of self-consciousness. When I say therefore that I am willing to regard such a robot as an "organism," I declare my willingness to consider it a kind of animal. And I have already agreed that I see no way to put a bound on the degree of intelligence such an organism could, at least in principle, attain.

I make these stipulations, as the lawyers would call them, not because I believe that what any reasonable observer would call a socialized robot is going to be developed in the "visible future" – I do not believe that – but to avoid the unnecessary, interminable, and ultimately sterile exercise of making a catalogue of what computers will and will not be able to do, either here and now or ever. That exercise would deflect us from the primary question, namely, whether there are objectives that are not appropriately assignable to machines.

If both machines and humans are socializable, then we must ask in what way the socialization of the human must necessarily be different from that of the machine. The answer is, of course, so obvious that it makes the very asking of the question appear ludicrous, if indeed not obscene. It is a sign of the madness of our time that this issue has to be addressed at all.

Every organism is socialized by the process of dealing with problems that confront it. The very biological properties that differentiate one species from another also determine that each species will confront problems different from those faced by any other. Every species will, if only for that reason, be socialized differently. The human infant, as many observers have remarked, is born prematurely, that is, in a state of utter helplessness. Yet the infant has biological needs which, if he is to survive at all, must be satisfied by others. Indeed, many studies of orphanages have shown that more than his merely elementary physical needs must be satisfied; an infant will die if he is fed and cleaned but not, from the very beginning of his life, fondled and caressed – if, in other words, he is not treated as a human being by other human beings.

A catastrophe, to use Erik Erikson's expression for it, that every human being must experience is his personal recapitulation of the biblical story of paradise. For a time the infant demands and is granted gratification of his every need, but is asked for nothing in return. Then, often after the infant has developed teeth and has bitten the breast that has fed him, the unity between him and his mother is broken. Erikson believes this universal human drama to be the ontogenetic contribution to the biblical saga of the Garden of Eden. So important is this period in the child's life that

> a drastic loss of accustomed mother love without proper substitution at this time can lead [under otherwise aggravating conditions] to acute infantile depression or to a mild but chronic state of mourning which may give a depressive undertone to the whole remainder of life. But even under the most favorable circumstances, this stage leaves a residue of a primary sense of evil and doom and of a universal nostalgia for a lost paradise.
>
> [These early stages] then, form in the infant the springs of the basic sense of trust and the basic sense of mistrust which remain the autogenic source of both primal hope and of doom throughout life.

Thus begins the individual human's imaginative reconstruction of the world. And this world, as I said earlier, is the repository of his subjectivity, the stimulator of his consciousness, and ultimately the constructor of the apparently external forces he is to confront all his life.

> As the child's radius of awareness, co-ordination, and responsiveness expands, he meets the educative patterns of his culture, and thus learns the basic modalities of human existence, each in personally and culturally significant ways ... *To get* ... means to receive and to accept what is given. This is the first social modality learned in life; and it sounds simpler than it is. For the groping and unstable newborn organism learns this modality only as it learns to regulate its organ systems in accordance with the way in which the maternal environment integrates its methods of child care....
>
> The optimum total situation implied in the baby's readiness to get what is given is his mutual regulation with a mother who will permit him to develop and coordinate his means of getting as she develops and co-ordinates her means of giving. ... The mouth and the nipple seem to be the mere centers of a general aura of warmth and mutuality which are enjoyed and responded to with relaxation not only by these focal organs, but by both total organisms. The mutuality of relaxation thus developed is of prime importance for the first experience of friendly otherness. One may say ... that in thus *getting what is given*, and in learning to *get somebody to do* for him what he wishes to have done, the baby also develops the necessary ego groundwork *to get to be* a giver.

What these words of Erikson's make clear is that the initial and crucial stages of human socialization implicate and enmesh the totality of two organisms, the child and its mother, in an inseparable mutuality of service to their deepest biological and emotional needs. And out of this problematic reunification of mother and child – problematic because it involves inevitably the trauma of separation – emerge the foundations of the human's knowledge of what it means to give and to receive, to trust and to mistrust, to be a friend and to have a friend, and to have a sense of hope and a sense of doom.

[...] [N]o term of a theory can ever be fully and finally understood. We may say the same thing about words generally, especially about such words as trust and friendship and hope and their derivatives. Erikson teaches us that such words derive their meanings from universal, primal human experiences, and that any understanding of them must always be fundamentally metaphoric. This profound truth also informs us that man's entire understanding of his world, since it is mediated by his language, must always and necessarily be bounded by metaphoric descriptions. And since the child "meets the educative patterns of his culture," as Erikson says, "and thus learns the basic modalities of human existence, each in personally and culturally significant ways", each culture, indeed, each individual in a culture, understands such words and language, hence the world, in a culturally and personally idiosyncratic way.

I could go on to describe the later stages of the socialization of the individual human, the effects of schooling, marriage, imprisonment, warfare, hates and loves, the experiences of shame and guilt that vary so radically among the cultures of man, and so on. But that could be of no help to anyone who is not already convinced that any "understanding" a computer may be said to possess, hence any "intelligence" that may be attributed to it, can have only the faintest relation to human understanding and human intelligence. We, however, conclude that however much intelligence computers may attain, now or in the future, theirs must always be an intelligence *alien* to genuine human problems and concerns.

Still, the extreme or hardcore wing of the artificial intelligentsia will insist that the whole man, to again use Simon's expression, is after all an information processor, and that an information-processing theory of man must therefore be adequate to account for his behavior in its entirety. We may agree with the major premise without necessarily drawing the indicated conclusion. We have already observed that a portion of the information the human "processes" is kinesthetic, that it is "stored" in his muscles and joints. It is simply not clear that such information, and the processing associated with it, can be represented in the form of computer programs and data structures at all.

It may, of course, be argued that it is in principle possible for a computer to simulate the entire network of cells that constitutes the human body. But that would introduce a theory of information processing entirely different from any which has so far been advanced. Besides, such a simulation would result in "behavior" on such an incredibly long time-scale that no robot built on such principles could possibly interact with human beings. Finally, there appears to be no prospect whatever that mankind will know enough neurophysiology within the next several hundred years to have the intellectual basis for designing such a machine. We may therefore dismiss such arguments.

There is, however, still another assumption that information-processing modelers of man make that may be false, and whose denial severely undermines their program: that there exists one and only one class of information processes, and that every member of that class is reducible to the kind of information processes exemplified by such systems as GPS and Schank-like language-understanding formalisms. Yet every human being has the impression that he thinks at least as much by intuition, hunch, and other such informal means as he does "systematically," that is by means such as logic. Questions like "Can a computer have original ideas? Can it compose a metaphor or a symphony or a poem?" keep cropping up. It is as if the folk wisdom knows the distinction between computer thought and the kind of thought people ordinarily engage in. The artificial intelligentsia, of course, do not believe there need be any distinction. They smile and answer "unproven."

Within the last decade or so, however, neurological evidence has begun to accumulate that suggests there may be a scientific basis to the folk wisdom. It has long been known that the human brain consists of two so-called hemispheres that appear, superficially at least, to be identical. These two halves, which we will call LH (Left Hemisphere) and RH (Right Hemisphere), have, however, quite distinct functions. In righthanded people – and for simplicity, we can restrict our discussion to them – the LH may be said, at least roughly, to control the right half of the body, and the RH the left half. (Actually, the connectivities are somewhat more complex, particularly between the two brain halves and the eyes, but I will not go into such details here.) Most importantly, the two halves of the brain appear to have two quite distinct modalities of thought. The LH thinks, so to speak, in an orderly, sequential, and, we might call it, logical fashion. The RH, on the other hand, appears to think in terms of holistic images. Language processing appears to be almost exclusively centered in the LH, for example, whereas the RH is deeply involved in such tasks as spatial orientation, and the production and appreciation of music.

The distinct functions of the hemispheres of the brain began to be dramatically illustrated by patients who, after suffering from extremely severe forms of epilepsy, had their two brain halves surgically separated. In normal people, the two hemispheres are connected by a part of the brain called the corpus callosum. When this is cut, no direct communication between the two halves remains possible. It was found that when a so-called split-brain patient's hands were visually hidden from him and he was given, say, a pencil in his left hand, he could not *say* what had been given to him, but he could *show* that it was a pencil by drawing a picture of it or by selecting a picture of a pencil from among pictures of many different objects. However, when the experiment was repeated, only with the right hand receiving the pencil, then he could say it was a pencil but could not produce or recognize its pictorial representation. In the first situation, the RH received the "image" of the pencil and was able to encode it into pictorial representations, but not into linguistic structures. In the second, the LH received the "image" of the pencil and was able to encode it linguistically, but not pictorially.

There is also considerable evidence, which I will not detail here, that the RH is essentially the seat of intuition, and that it thinks quite independently of the LH. One way of characterizing intuitive thought is to say that, although it is logical, the standards of evidence it uses to make judgments are very different from the standards we normally associate with logical thought. In ordinary discourse, for example, when we say that two things are the same, we mean that they are identical in almost every respect; the standard of evidence we demand to justify such a judgment is extremely demanding. But when we construct a metaphor, e.g., the overseas Chinese are the Jews of the Orient, we pronounce two things to be the same in a very different sense. Metaphors are simply not logical; when taken literally, they are patently absurd. The RH, in other words, has criteria of absurdity that are far different from those of the logical LH.

The history of man's creativity is filled with stories of artists and scientists who, after working hard and long on some difficult problem, consciously decide to "forget" it, in effect, to turn it over to their RH. After some time, often with great suddenness and totally unexpectedly, the solution to their problem announces itself to them in almost complete form. The RH appears to have been able to overcome the most difficult logical and systematic problems by, I would conjecture, relaxing the rigid standards of thought of the LH. Given the looser standards the RH employs, it was perhaps able to design thought experiments which the LH simply could not, because of its rigidity, conceive. The RH is thus able to hit upon solutions which could then, of course, be recast into strictly logical terms by the LH. We may conjecture that in children the communication channel between the two brain halves is wide open; that is, that messages pass between the two halves quite freely. That may be why children are so incredibly imaginative; e.g., for them a cigar box is an automobile one moment and a house the next. In adults, the channel has been severely narrowed – whether by education or by physiological maturational processes or by both, I cannot guess. But it is clearly more open during the dream state. I may also conjecture that psychoanalysis, quite apart from its function as psychotherapy, trains people in the use of the channel. In psychoanalysis one learns, in Theodore Reik's happy phrase, to listen with the third ear, to attend, that is, to what the unconscious is "saying." Perhaps the various meditative disciplines serve the same purpose.

These are clearly conjectures, from which we are not entitled to draw any conclusions about how either humans or computers process information. Even as a mere possibility, however, they do raise a serious question about the universality of the mode of information processing we normally associate with logical thought and with computer programs.

That the right hemisphere of the brain is, loosely speaking, the "seat of intuition" is a hypothesis in favor of which evidence appears to be accumulating. Neither philosophers nor psychologists have yet been sufficiently persuaded by the existing evidence to confidently incorporate this hypothesis into their theories of mind. But this much is firmly established: the two hemispheres of the human brain think independently of one another; they think simultaneously; and they think in modes different from one another. Furthermore, we can say something about these two distinct modes.

The great mathematician Henri Poincaré, in his celebrated essay *Mathematical Creation,* wrote:

> The conscious self is narrowly limited, and as for the subliminal self we know not its limitations ... calculations ... must be made in the ... period of conscious work, that which follows the inspiration, that in which one verifies the results of this inspiration and deduces their consequences. The rules of these calculations are strict and complicated. They require discipline, attention, will, and therefore consciousness. In the subliminal self, on the contrary, reigns what I should call liberty, if we might give this name to the simple absence of discipline ... the privileged unconscious phenomena, those susceptible of becoming conscious, are those which, directly or indirectly, affect most profoundly our emotional sensibility ... The role of this unconscious work in mathematical invention appears to me incontestable, and traces of it would be found in other cases where it is less evident.

Of course, Poincaré, writing at the beginning of the twentieth century, knew nothing of the findings of the now-active brain researchers. And we are jumping to a conclusion when we identify what he calls the conscious and the subliminal selves with the left and right hemispheres of the brain, respectively. But our assertion here is that there are two distinct modes of human thought that operate independently and simultaneously. And that assertion Poincaré supports.

A most highly respected scientist who is now working, the psychologist Jerome Bruner, writes on this same topic from a slightly different perspective (recall that the right hand corresponds to the left hemisphere and the left hand to the right, or "intuitive," hemisphere):

> As a right-handed psychologist, I have been diligent for fifteen years in the study of the cognitive processes: how we acquire, retain, and transform knowledge of the world in which each of us lives – a world in part 'outside' us, in part 'inside.' The tools I have used have been those of the scientific psychologist studying perception, memory, learning, thinking, and (like a child of my times) I have addressed my inquiries to the laboratory rat as well as to human beings. At times, indeed, I have adopted the role of the clinician and carried out therapy with children. ... There have been times when, somewhat discouraged by the complexities of the psychology of knowing, I have sought to escape through neurophysiology, to discover that the neurophysiologist can help only in the degree to which we can ask intelligent psychological questions of him.
>
> One thing has become increasingly clear in pursuing the nature of knowing. It is that the conventional apparatus of the psychologist – both his instruments

> of investigation and the conceptual tools he uses in the interpretation of his data – leaves one approach unexplored. It is an approach whose medium of exchange seems to be the metaphor paid out by the left hand. It is a way that grows happy hunches and 'lucky' guesses, that is stirred into connective activity by the poet and the necromancer looking sidewise rather than directly. Their hunches and intuitions generate a grammar of their own – searching out connections, suggesting similarities, weaving ideas loosely in a trial web...
>
> [The psychologist] too searches widely and metaphorically for his hunches. He reads novels, looks at and even paints pictures, is struck by the power of myth, observes his fellow men intuitively and with wonder. In doing so, he acts only part-time like a proper psychologist, racking up cases against the criteria derived from hypothesis. Like his fellows, he observes the human scene with such sensibility as he can muster in the hope that his insight will be deepened. If he is lucky or if he has subtle psychological intuition, he will from time to time come up with hunches, combinatorial products of his metaphoric activity. If he is not fearful of these products of his own subjectivity, he will go so far as to tame the metaphors that have produced the hunches, tame them in the sense of shifting them from the left hand to the right hand by rendering them into notions that can be tested. It is my impression from observing myself and my colleagues that the forging of metaphoric hunch into testable hypothesis goes on all the time.

That, of course, is my impression as well. Here Bruner speaks explicitly of the left hand, that is, the right hemisphere of the brain, as the artistic, the intuitive, and so on, and of the right hand, the left brain hemisphere, as the "conventional apparatus of the psychologist," and he speaks of the inadequacy of his "conceptual tools."

We learn from the testimony of hundreds of creative people, as well as from our own introspection, that the human creative act always involves the conscious interpretation of messages coming from the unconscious, the shifting of ideas from the left hand to the right, in Bruner's phrase.

The unconscious is, of course, unconscious. It is like a seething, stormy sea within us. Its waves lap on the borders of our consciousness. And what we learn from it or about it, we construct from inferences we make about the meanings of the swells and surges, the breakers and ripples that wash the fringes of our consciousness. Occasionally we wander more deeply into the surf, as when we are in that semi-hypnagogic trance that divides sleep from wakefulness. But then we experience only chaos. Our thought modalities are maximally confused. And if we rip ourselves into waking, we cannot tell, we cannot translate or transform into linguistic modalities, what we had thought.

Does not the undoubted reality of this confusion, when placed alongside all the other available evidence to which I have alluded, lend weight to the altogether plausible conjecture that the forms of information manipulated in the right hemisphere of the brain, as well as the corresponding information processes, are simply different from those of the left hemisphere? And may it not be that we can in principle come to know those strange information forms and processes only in terms that are fundamentally irrelevant to the kind of understanding we seek? When, in the distant future, we come to know in detail how the brain functions on the neurophysiological level, we will, of course, be able to give an ultimately reductionist account of the functioning of the right hemisphere. But that would not be understanding in the sense we mean here, anymore than detailed knowledge of the electrical behavior of a running computer is, or even leads to, an understanding of the program the computer is running. On the other hand, a higher-level account of the functioning of the right hemisphere may always miss its most essential features, namely, those that differentiate it from the functioning of the left hemisphere. For we are constrained by our left hemisphere thought modalities to always interpret messages coming from the right in left-hemisphere terms.

Perhaps the LH modality of thought is GPS-like, which is to say only that perhaps it can in principle be somehow formulated (not that GPS is even a candidate for a possible formulation). Perhaps it converts a problem like:

> Tom has twice as many fish as Mary has guppies. If Mary has three guppies, how many fish does Tom have?

into its own terms, for example into

$$x = 2y; y = 3,$$

and solves it using information processing and symbol-manipulation techniques characteristic of GPS-like "thought." But it is then not possible for such a mechanism to have any idea of what fishes and guppies are, or of what it can mean to be a boy named Tom, and so on. Nor can the symbolic representation of the given problem be reconverted into the original problem statement. But human problem-solving, perhaps even of the apparently most routine and mechanical variety, involves both left and right modes of thought. And certainly, direct human communication crucially involves the two hemispheres.

It is much too easy, especially for computer scientists, to be hypnotized by the "fact" that linguistic utterances are representable as linear strings of symbols. From this "fact" it is easy to deduce that linguistic communication is entirely a left-hemisphere affair. But human speech also has melody, and its song communicates as well as its libretto. Music is the province of the right hemisphere, as is the appreciation of gestures. As for written communication, its function is surely, at least in large part, to stimulate and excite especially the auditory imaginations of both the writer and the reader.

We may never know whether the conjecture that a part of us thinks in terms of symbolic structures that can be only sensed but not usefully explicated is true or false. Scientists, of course, abhor hypotheses that appear not to be falsifiable. Yet it may be that, under some profound conception of truth, the hypothesis is true. Perhaps it helps to explain why we remain lifelong strangers to ourselves and to each other, why every word in our lexicon is enveloped in at least some residual mystery, and why every attempt to solve life's problems by entirely rational means always fails.

But the inference that I here wish to draw from my conjecture is that, since we cannot know that it is false any more than that it is true, we are not entitled to the hubris so bombastically exhibited by the artificial intelligentsia. Even calculating reason compels the belief that we must stand in awe of the mysterious spectacle that is the whole man – I would even add, that is the whole ant.

There was a time when physics dreamed of explaining the whole of physical reality in terms of one comprehensive formalism. Leibnitz taught that if we knew the position and velocity of every elementary particle in the universe, we could predict the universe's whole future course. But then Werner Heisenberg proved that the very instruments man must use in order to measure physical phenomena disturb those phenomena, and that it is therefore impossible in principle to know both the exact position and the velocity of even a single elementary particle. He did not thereby falsify Leibnitz's conjecture. But he did show that its major premise was unattainable. That, of course, was sufficient to shatter the Leibnitzian dream. Only a little later, Kurt Gödel exposed the shakiness of the foundations of mathematics and logic itself by proving that every interesting formal system has some statements whose truth or falsity cannot be decided by the formal means of the system itself, in other words, that mathematics must necessarily be forever incomplete. It follows from this and others of Gödel's results that "The human mind is incapable of formulating (or mechanizing) all its mathematical intuitions. I.e.: If it has succeeded in formulating some of them, this very fact yields new intuitive knowledge."

Both Heisenberg's so-called uncertainty principle and Gödel's incompleteness theorem sent terrible shock-waves through the worlds of physics, mathematics, and philosophy of science. But no one stopped working. Physicists, mathematicians, and philosophers more or less gracefully accepted the undeniable truth that there are limits to how far the world can be comprehended in Leibnitzian terms alone. Much too much has already been made of the presumed implications of Heisenberg's and Gödel's results for artificial intelligence. I do not wish to contribute to that discussion here. But there is a sense in which psychology and artificial intelligence may usefully follow the example of the new-found humility of modern mathematics and physics: they should recognize that "while the constraints and limitations of logic do not exert their force on the things of the world, they do constrain and limit what are to count as defensible descriptions and interpretations of things." Were they to recognize that, they could then take the next liberating step of also recognizing that truth is not equivalent to formal provability.

The lesson I have tried to teach here is not that the human mind is subject to Heisenberg uncertainties – though it may be – and that we can therefore never wholly comprehend it in terms of the kinds of reduction to discrete phenomena Leibnitz had in mind. The lesson here is rather that the part of the human mind which communicates to us in rational and scientific terms is itself an instrument that disturbs what it observes, particularly its voiceless partner, the unconscious, between which and our conscious selves it mediates. Its constraints and limitations circumscribe what are to constitute rational – again, if you will, scientific – descriptions and interpretations of the things of the world. These descriptions can therefore never be whole, any more than a musical score can be a whole description or interpretation of even the simplest song.

But, and this is the saving grace of which an insolent and arrogant scientism attempts to rob us, we come to know and understand not only by way of the mechanisms of the conscious. We are capable of listening with the third ear, of sensing living truth that is truth beyond any standards of provability. It is *that* kind of understanding, and the kind of intelligence that is derived from it, which I claim is beyond the abilities of computers to simulate.

We have the habit, and it is sometimes useful to us, of speaking of man, mind, intelligence, and other such universal concepts. But gradually, even slyly, our own minds become infected with what A. N. Whitehead called the fallacy of misplaced concreteness. We come to believe that these theoretical terms are ultimately interpretable as observations, that in the "visible future" we will have ingenious instruments capable of measuring the "objects" to which these terms refer. There is, however, no such thing as mind; there are only individual minds, each belonging, not to "man," but to individual human beings. I have argued that intelligence cannot be measured by ingeniously constructed meter sticks placed along a one-dimensional continuum. Intelligence can be usefully discussed only in terms of domains of thought and action. From this I derive the conclusion that it cannot be useful, to say the least, to base serious work on notions of "how much" intelligence may be given to a computer. Debates based on such ideas – e.g., "Will computers ever exceed man in intelligence?" – are doomed to sterility.

I have argued that the individual human being, like any other organism, is defined by the problems he confronts. The human is unique by virtue of the fact that he must necessarily confront problems that arise from his unique biological and emotional needs. The human individual is in a constant state of becoming. The maintenance of that state, of his humanity, indeed, of his survival, depends crucially on his seeing himself, and on his being seen by other human beings, as a human being. No other organism, and certainly no computer, can be made to confront genuine human problems in human terms. And, since the domain of human intelligence is, except for a small set of formal problems, determined by man's humanity, every other intelligence, however great, must necessarily be alien to the human domain.

I have argued that there is an aspect to the human mind, the unconscious, that cannot be explained by the information-processing primitives, the elementary information processes, which we associate with formal thinking, calculation, and systematic rationality. Yet we are constrained to use them for scientific explanation, description, and interpretation. It behoves us, therefore, to remain aware of the poverty of our explanations and of their strictly limited scope. It is wrong to assert that any scientific account of the "whole man" is possible. There are some things beyond the power of science to fully comprehend.

The concept of an intelligence alien to certain domains of thought and action is crucial for understanding what are perhaps the most important limits on artificial intelligence. But that concept applies to the way humans relate to one another as well as to machines and their relation to man. For human socialization, though it is grounded in the biological constitution common to all humans, is strongly determined by culture. And human cultures differ radically among themselves. Countless studies confirm what must be obvious to all but the most parochial observers of the human scene: "The influence of culture is universal in that in some respects a man learns to become like all men; and it is particular in that a man who is reared in one society learns to become in some respects like all men of his society and not like those of others." The authors of this quotation, students of Japanese society who lived among the Japanese for many years, go on to make the following observations:

> In normal family life in Japan there is an emphasis on interdependence and reliance on others, while in America the emphasis is on independence and self-assertion ... In Japan the infant is seen more as a separate biological organism who from the beginning, in order to develop, needs to be drawn into increasingly interdependent relations with others. In America, the infant is seen more as a dependent biological organism who, in order to develop, needs to be made increasingly independent of others.
>
> The Japanese baby seems passive, and he lies quietly with occasional unhappy vocalizations, while his mother, in her care, does more lulling, carrying, and rocking of her baby. She seems to try to soothe and quiet the child, and to communicate with him physically rather than verbally. On the other hand, the American infant is more active, happily vocal, and exploring of his environment, and his mother, in her care, does more looking at and chatting to her baby. She seems to stimulate the baby to activity and to vocal response. It is as if the American mother wanted to have a vocal, active baby, and the Japanese mother wanted to have a quiet, contented baby. In terms of styles of caretaking of the mothers in the two cultures, they get what they apparently want ... a great deal of cultural learning has taken place by three-to-four months of age ... babies have learned by this time to be Japanese and American babies in relation to the expectations of their mothers concerning their behavior.
>
> [Adult] Japanese are more 'group' oriented and interdependent in their relations with others, while Americans are more 'individual' oriented and independent ... Japanese are more self-effacing and passive in contrast to Americans, who appear more self-assertive and aggressive ... Japanese are more sensitive to, and make conscious use of, many forms of nonverbal communication in human relations through the medium of gestures and physical proximity in comparison with Americans, who predominantly use verbal communication within a context of physical separateness.
>
> If these distinct patterns of behavior are well on the way to being learned by three-to-four months of age, and if they continue over the life span of the person, then there are very likely to be important areas of difference in emotional response in people of one culture when compared with those in another. Such differences are not easily subject to conscious control and, largely out of awareness, they accent and color human behavior. These differences ... can also add to bewilderment and antagonism when people try to communicate across the emotional barriers of culture.

Such profound differences in early training crucially affect the entire societies involved. And they are, of course, transmitted from one generation to the next and thus perpetuated. They must necessarily also help determine what members of the two societies know about their worlds, what are to be taken as "universal" cultural norms and values, hence what in each culture is and is not to be counted as fact. They determine, for example (and this is particularly relevant to the contrast between Japanese and American social norms), what are private as opposed to public conflicts, and hence what modes of adjudication are appropriate to the defense of what human interests. The Japanese traditionally prefer to settle disputes, even those for which relief at law is statutorily available, by what Westerners would see as informal means. Actually, these means are most often themselves circumscribed by stringent ritualistic requirements that are nowhere explicitly codified but are known to every Japanese of the appropriate social class. This sort of knowledge is acquired with the mother's milk and through the whole process of socialization that is itself so intimately tied to the individual's acquisition of his mother tongue. It cannot be learned from books; it cannot be explicated in any form but life itself.

An American judge, therefore, no matter what his intelligence and fairmindedness, could not sit in a Japanese family court. His intelligence is simply alien to the problems that arise in Japanese culture. The United States Supreme Court actively recognized this while it still had jurisdiction over distant territories. For example, in the case of Diaz v. Gonzales, which was originally tried in Puerto Rico, the court refused to set aside the judgment of the court of original jurisdiction, that is, of the native court. Justice Oliver W. Holmes, writing the opinion of the Court, stated,

> This Court has stated many times the deference due to understanding of the local courts upon matters of purely local concern. This is especially true when dealing with the decisions of a Court inheriting and brought up in a different system from that which prevails here. When we contemplate such a system from the outside it seems like a wall of stone, every part even with all the others, except so far as our own local education may lead us to see subordinations to which we are accustomed. But to one brought up within it, varying emphasis, tacit assumptions, unwritten practices, a thousand influences gained only from life, may give to the different parts wholly new values that logic and grammar never could have got from the books.

Every human intelligence is thus alien to a great many domains of thought and action. There are vast areas of authentically human concern in every culture in which no member of another culture can possibly make responsible decisions. It is not that the outsider is unable to decide at all – he can always flip coins, for example – it is rather that the *basis* on which he would have to decide must be inappropriate to the context in which the decision is to be made.

What could be more obvious than the fact that, whatever intelligence a computer can muster, however it may be acquired, it must always and necessarily be absolutely alien to any and all authentic human concerns? The very asking of the question, "What does a judge (or a psychiatrist) know that we cannot tell a computer?" is a monstrous obscenity. That it has to be put into print at all, even for the purpose of exposing its morbidity, is a sign of the madness of our times.

Computers can make judicial decisions, computers can make psychiatric judgments. They can flip coins in much more sophisticated ways than can the most patient human being. The point is that they *ought* not be given such tasks. They may even be able to arrive at "correct" decisions in some cases – but always and necessarily on bases no human being should be willing to accept.

There have been many debates on "Computers and Mind." What I conclude here is that the relevant issues are neither technological nor even mathematical; they are ethical. They cannot be settled by asking questions beginning with "can." The limits of the applicability of computers are ultimately statable only in terms of oughts. What emerges as the most elementary insight is that, since we do not now have any ways of making computers wise, we ought not now to give computers tasks that demand wisdom.

What computers mean for man and society

Herbert A. Simon

Energy and information are two basic currencies of organic and social systems. A new technology that alters the terms on which one or the other of these is available to a system can work on it the most profound changes. At the core of the industrial revolution, which began nearly three centuries ago, lay the substitution of mechanical energy for the energy of man and animal. It was this revolution that changed a rural subsistence society into an urban affluent one and touched off a chain of technological innovations that transformed not only production but also transportation, communication, warfare, the size of human populations, and the natural environment.

It is easy, by hindsight, to see how inexorably these changes followed one another, how "natural" a consequence, for example, suburbia was of cheap, privately owned transportation. It is a different question whether foresight could have predicted these chains of events or have aided in averting some of their more undesirable outcomes. The problem is not that prophets were lacking – they have been in good supply at almost all times and places. Quite the contrary, almost everything that has happened, and its opposite, has been prophesied. The problem has always been to pick and choose among the embarrassing riches of alternative projected futures; and in this, human societies have not demonstrated any large foresight. Most often we have been constrained to anticipate events just a few years before their occurence, or even while they are happening, and to try to deal with them, as best we can, as they are engulfing us.

We are now in the early stages of a revolution in processing information that shows every sign of being as fundamental as the earlier energy revolution. Perhaps we should call it the third information revolution. (The first produced written language, and the second, the printed book.) This third revolution, which began more than a century ago, includes the computer but many other things as well. The technology of information comprises a vast range of processes for storing information, for copying it, for transmitting it from one place to another, for displaying it, and for transforming it.

Photography, the moving picture, and television gave us, in the course of a century, a whole new technology for storing and displaying pictorial information. Telegraphy, the telephone, the phonograph, and radio did the same for storing and transmitting auditory information. Among all of these techniques, however, the computer is unique in its capacity for manipulating and transforming information and hence in carrying out, automatically and without human intervention, functions that had previously been performable only by the human brain.

As with the energy revolution, the consequences of the information revolution spread out in many directions. First, there are the economic consequences that follow on any innovation that increases human productivity. As we shall see, these are perhaps the easiest effects of technological change to predict. Second, there are consequences for the nature of work and of leisure – for the quality of life. Third, the computer may have special consequences for privacy and individual liberty. Fourth, there are consequences for man's view of himself, for his picture of the universe and of his place and goals in it. In each of these directions, the immediate consequences are, of course, the most readily perceived. (It was not hard to foresee that Newcomen's and Watt's engines would change the economics of mining in deep pits.) It is far more difficult to predict what indirect chains of effects these initial impacts will set off, for example, the chain that reaches from the steam engine through the internal-combustion engine to the automobile and the suburb.

Prediction is easier if we do not try to forecast in detail the time path of events and the exact dates on which particular developments are going to occur, but to focus, instead, upon the steady state toward which the system is tending. Of course, we are not so much interested in what is going to happen in some vague and indefinite future as we are in what the next generation or two holds for us. Hence, a generation is the time span with which I shall be concerned.

My discussion will be divided into five parts, the last four corresponding to domains of prediction: economics, the nature of work and leisure, social consequences, and how men and women view themselves. These essays in prediction need to be preceded, however, by some analysis of the computer itself, and particularly its capabilities and potential in the area that is usually called artificial intelligence. This subject is taken up in the next section.

Computer capabilities

The computer is a device endowed with powers of utmost generality for processing symbols. It is remarkable not only for its capabilities but also for the simplicity of its underlying processes and organization. Of course, from a hardware standpoint it is not simple at all but is a highly sophisticated electronic machine. The simplicity appears at the level of the elementary information processes that the hardware enables it to perform, the organization for execution and control of those processes, and the programming languages in terms of which the control of its behavior is expressed. A computer can read symbols from an external source, output symbols to an external destination, store symbols in one or more memories, copy symbols, rearrange symbols and structures of symbols, and react to symbols conditionally – that is, follow one course of action or another, depending on what symbols it finds in memory or in its input devices. The most general symbol-manipulating system that has been defined, the so-called Turing machine, requires no broader capabilities than these. The important limits on the powers of a computer are limits on the sizes of its memories and the speed of its elementary processes, and not on the generality of those processes.

There is great dispute among experts as to what the generality of the computer implies for its ability to behave intelligently. There is also dispute as to whether the computer, when it is behaving more or less intelligently, is using processes similar to those employed by an intelligent human being, or quite different processes. The views expressed here will reflect my own experience in research with computers and my interpretation of the scientific literature. First, no limits have been discovered to the potential scope of computer intelligence that are not also limits on human intelligence. Second, the elementary processes underlying human thinking are essentially the same as the computer's elementary information processes, although modern fast computers can execute these processes more rapidly than can the human brain. In the past, computer memories, even in large computers, have probably not been as capacious as human memory, but the scale of available computer memories is increasing rapidly, to the point where memory size may not be much longer an effective limit on the capacity of computers to match human performance. Any estimate of the potential of the computer in the near or distant future depends on one's agreement or disagreement with these assumptions.

One common objection to the beliefs just expressed is that "computers can only do what you program them to do." That is correct. The behavior of a computer at any specific moment is completely determined by the contents of its memory and the symbols that are input to it at that moment. This does not mean that the programmer must anticipate and prescribe in the program the precise course of its behavior. A program is not a scenario; it is a strategy of action, and what actions actually transpire depends on the successive states of the machine and its inputs at each stage of the process – neither of which need be envisioned in advance either by the programmer or by the machine. A problem-solving program applied to a particular puzzle situation does not prescribe all the steps to solve that puzzle; it prescribes a selective search strategy that, when followed, may lead the computer to discover a path to a solution. But selective search, under the guidance of strategies, is also the process that people use to solve puzzles.

Of course humans, through the processes called learning, can improve their strategies by experience and instruction. By the same token, computers can be, and to some extent have been, provided with programs (strategies) for improving their own strategies. Since a computer's programs are stored in the same memories as data, it is entirely possible for programs to modify themselves – that is, to learn.

Probably the most fundamental differences between today's computers and the human information-processing system have to do with the input organs that provide the interface between the system and its environment. Simulating the capabilities of human eyes and ears has proved to be a much more difficult task than simulating the thinking processes that go on in the central nervous system. Computer capabilities in both visual and auditory domains, and particularly the former, fall far short of human capabilities.

Over the past two decades a moderate amount of work has been carried on in the field usually called artificial intelligence to explore the potentialities of the computer that have been outlined above. Some of this research has been aimed at programming computers to do things which, if done by a person, would be regarded as intelligent. Another part of the research has been directed at simulating not only the human capabilities but also the processes that human beings use in exercising these capabilities. The considerable progress that has been made in understanding the nature both of artificial and of human intelligence has hardly begun to translate itself into applications, and has been reflected to only a small degree in the actual practical uses of computers. Artificial intelligence research has had an impact upon the search algorithms that are used to solve large combinatorial problems, it is on the verge of practical application in the realm of medical diagnosis, and it has had an important influence upon certain computer programming techniques (for example, list processing). But its main significance for practical affairs lies in the future.

How, then, have computers actually been used to date? At present, computers typically spend most of their time in two main kinds of tasks: carrying out large-scale engineering and scientific calculations and keeping the financial, production, and sales records of business firms and other organizations. Although precise statistics are not available, it would be safe to estimate that 95 per cent of all computing power is allocated to such jobs. Now these tasks belong to the horseless-carriage phase of computer development. That is to say, they consist in doing things rapidly and automatically that were being done slowly and by hand (or by desk calculator) in the pre-computer era.

Such uses of computers do not represent new functions but only new ways of performing old functions. Of course, by greatly lowering the cost of performing them, they encourage us to undertake them on a larger scale than before. The increased analytic power provided by computers has probably encouraged engineers to design more complex structures (for example, some of the very tall new office buildings that have gone up in New York and Chicago) than they would have attempted if their analytic aids were less powerful. Moreover, by permitting more sophisticated analyses to be carried out in the design process, they have also brought about significant cost reductions in the designs themselves. In the same way, the mechanization of business record-keeping processes has facilitated the introduction of improved controls over inventories and cash flows, with resulting savings in costs. Thus, the computer not only reduces the costs of the information-processing operations that it automates but also contributes to the productivity of the activities themselves.

The remaining five per cent of computer uses are more sophisticated.

Let us consider two different ways in which a computer can assist an engineer in designing electric motors. On the one hand, the engineer can design the motor using conventional procedures, then employ the computer to analyze the prospective operation of the design – the operating temperature, efficiency, and so on. On the other hand, the engineer can provide the computer with the specifications for the motor, leaving to the computer the task of synthesizing a suitable design. In the second, but not the first, case the computer, using various heuristic search procedures, actually discovers, decides upon, and evaluates a suitable design. In the same way, the role of the computer in managing inventories need not be limited to record-keeping. The computer program may itself determine (on the basis of usage) when items should be reordered and how large the orders should be. In these and many other situations, computers can provide not only the information on which decisions are made but can themselves make the decisions. Process-control computers, in automated or semi-automated manufacturing operations, play a similar role in decision-making. Their programs are decision strategies which, as the system's variables change from moment to moment, retain control over the ongoing process.

It is the capability of the computer for solving problems and making decisions that represents its real novelty and that poses the greatest difficulties in predicting its impact upon society. An enormous amount of research and developmental activity will have to be carried out before the full practical implications of this capability will be understood and available for use. In the single generation that modern computers have been in existence, enough basic research has been done to reveal some of the fundamental mechanisms. Although one can point to a number of applications of the computer as decision-maker that are already 20 or 25 years old, development and application on a substantial scale have barely begun.

Economic effects of computers

The direct economic effects of introducing computers as numerical calculators and decision-makers are like those of introducing any new form of capital that raises productivity and also improves the quality of the product. The computer (its hardware together with the associated system-programming costs) represents an investment in a capital-intensive, labour-saving device that has to justify itself, in competition with other possible forms of investment, through savings in clerical and other personnel costs together with the improvements it brings about in organizational decisions.

When the main motive of introducing the computer is to mechanize existing clerical operations – in the actuarial department of an insurance firm, say, or the accounting department of a manufacturing concern – then its main economic advantage stems from the reduction in clerical costs. When it is introduced to mechanize decision processes – engineering design, for example, or control of stock or cash inventories – then its direct effect shows up as some form of productivity increase in the organization's operations. In either case, there is nothing special about the computer that distinguishes it, in its economic effects, from any other capital investment. Any such investment can be expected to have a direct effect upon employment in the organizational components where it is introduced. When the accounting system is mechanized, fewer clerks and bookkeepers are needed, else there would be no economic motivation for mechanizing. Of course, if part of the motivation for the change is to improve the quality of the system's output, the operation may be expanded, and the net reduction in personnel may be smaller than would be estimated solely from the increase in efficiency. If there is sufficient elasticity of demand for the activity, personnel may actually increase.

The most important question, however, is what the reduction in personnel at the point of impact means for the total level of employment in the economy. Again, this is a general economic issue – of technological unemployment – that does not depend on any special properties of computers. They are simply one among the many labor-saving devices that have been appearing since the beginning of the Industrial Revolution (and before).

Both standard economic analysis and a large body of empirical evidence demonstrate that there is no relation, positive or negative, between the technological sophistication of an economy and the level of employment it maintains. From a systems standpoint, a cost reduction in any part of the system releases resources that can be employed to increase the output of goods and services elsewhere in the system. At any level of employment, from 0 to 100 per cent, the total revenue received by wage earners and owners of capital and land as wages, interest, and rent is just sufficient to purchase the total bundle of goods and services that is produced. Economists sometimes disagree as to why economies do not always operate at or near full employment, but they are unanimous in agreeing that the reason is not that they produce more than they can consume. (Even Marxists agree with this proposition, although they argue that full employment cannot be maintained within the institutions of capitalism.)

An even stronger statement can be made about the systems effects of cost-saving technological innovations. We usually describe devices like computers (and most other machinery) as labor-saving because they require a lower ratio of labor to capital than the methods they displace. But if we measure savings relative to output, they are usually both labor-saving and capital-saving. That is to say, a smaller capital investment per passenger mile is required to transport people by jet plane than to transport them by ox cart. Similarly, a smaller capital investment per multiplication is required if a large modern computer is used to do the arithmetic than if it is done on a desk calculator or with pencil and paper. (Do not forget to include the capital cost of the desk at which the clerk sits and the heated or air-conditioned room in which he or she works.) Now it is easy to show, for economic equilibrium and under reasonable assumptions about the supply of capital, that the introduction of capital-intensive, cost-saving innovations will raise the level of real wages and increase the fraction of the total revenue that goes to wages. This prediction from economic theory is amply supported by the histories of the industrialized economies over the present century. As productivity has increased (mainly as a consequence of technological innovation), real wages have steadily risen, as has labor's share in the total national product.

Now the rate of technological change depends both upon the rate of discovery of new innovations and upon the availability of capital to turn them into bricks and steel (or wire and glass). In this process, computers compete with other forms of technology for the available capital. Hence, the process of computerization is simply a part, currently an important part, of the general process of technological change. It might be described, paraphrasing Clausewitz, as "a continuation of the Industrial Revolution by other means".

In taking this very global and bird's eye view of the economics of mechanization, we should not ignore the plight of the worker who is displaced by the computer. His plight is often genuine and serious, particularly if the economy as a whole is not operating near full employment, but even if it is, Society as a whole benefits from increased productivity, but often at the expense of imposing transient costs on a few people. But the sensible response to this problem is not to eschew the benefits of change; it is rather to take institutional steps to shift the burdens of the transition from the individual to society. Fortunately, our attitudes on these questions appear to be maturing somewhat, and our institutional practices improving, so that the widespread introduction of the computer into clerical operations over the past generation has not called forth any large-scale Ludditism. In fact, during the depression that we are currently experiencing, in contrast to some earlier ones, technology has not been accused as the villain.

Effects on the nature of work

We see that, so far as economic effects are concerned, the computer simply provides a particular path toward higher productivity through industrialization. Whatever benefits it produces, it produces in this way; whatever problems it creates, it creates as other capital-intensive innovations do. We must be careful, however, not to evaluate social change solely in terms of its impact on wages and employment. Of equal importance are the effects it may have on the workplace, and even on leisure. Today we frequently hear the claim that computers and automation dehumanize work and that dehumanization, in turn, causes alienation from work and society. These charges have been laid not only against contemporary developments in automation but against the whole process of industrialization. They were stated eloquently in the *Communist Manifesto* more than a century ago and by numerous social critics before and since. There has been a new surge of concern with the alienation issue in the past 10 years.

Three questions need to be asked about alienation. First, how much alienation is there – is there evidence that alienation has been increased by computers and automation or, for that matter, by other forms of industrialization and mechanization? Second, in what ways is the nature of work, and the satisfactions derivable from it, changed by automation of the workplace? Third, as automation eliminates certain kinds of jobs in favor of others, what are the net effects upon the profile of jobs in the economy – are the jobs that are eliminated, on balance, more or less satisfying than the new ones that are created to replace them?

Objective data on national trends in job satisfaction are available only for about the last 20 years. About 15 national surveys have been conducted since 1958 by professional polling organizations that included questions on job satisfaction. Although 20 years is a short time, it does cover almost the whole period of the introduction of computers; hence these data should help answer the question before us. The polls provide absolutely no evidence for a decrease in job satisfactions over this period. If alienation has been increased by automation, the increase somehow does not show up in answers by workers to questions about their attitudes toward their jobs. Notice that the polls do not show that workers are enthusiastic about their jobs, only that they do not seem to like them less today than they did in 1958.

Unfortunately, comparable data are not available to measure the longer trends in job satisfaction over the whole past two centuries or so of industrialization. Perhaps even if computers and automation do not intensify alienation, they confirm and complete a loss of satisfactions that was produced by the rise of the factory system. The answer to that question must be mainly speculative. Clayre, however, recently threw some interesting light on it by examining the attitudes toward work expressed in preindustrial folk literature and popular ballads. He finds few indications of a Golden Age in which work was generally regarded as pleasurable and satisfying. He concludes that, in general, daily work was the same burdensome necessity for peasants and craftsmen as it is for factory workers and clerks; and that life's satisfactions and pleasure were mainly sought, then as now, in leisure, not work.

Perhaps, however, we should not try to detect alienation in this indirect way but should look directly at the workplaces where computers have been introduced, in order to see how they have changed the nature of work and its environment. Sizable differences have been found in worker satisfaction among blue collar workers in different kinds of factories, some of the important variables being job variety and worker control over the timing of work. It is not the case, however, that the most advanced forms of industrialization and automation produce the most tedious and restrictive jobs. On the contrary, those forms of work organization that appear to have been most alienating – typified by the auto assembly line or large-scale hand assembly operations – are declining in importance relative to other forms of mechanization. Blauner, for example, studied four industries in depth: printing, a traditional craft industry; textiles, a machine-tending industry; automobile assembly, highly mechanized with highly specialized jobs; and a highly automated continuous-process chemical manufacturing industry. He found few indications of alienation in printing and chemicals, considerably more in textiles, and most of all in automobile assembly. The industry that best typifies modern automation – chemicals – was substantially less alienating than the two that typify older kinds of mechanization.

If we look at office automation, we see that, here too, the kinds of jobs that are displaced tend to be those that are most repetitive and restricting. Whisler, who studied about 20 companies in the insurance industry, found that computerization had produced only small and conflicting changes in the nature of clerical and supervisory jobs. The new jobs placed greater demands on the employees for accuracy and reliability in performance, but they were not generally perceived as being significantly more or less pleasant or more or less boring than before. And, perhaps most important of all, whatever effects were produced were small effects. Automation and computerization do not appear to change the nature of work in a fundamental way.

Again we must look not just at immediate impact but at system effects. Factory and office automation are labor-saving technologies. The jobs they eliminate are mostly jobs that were already relatively routine. Therefore, when we look at the impact on the labor force as a whole, we expect to see automation bringing about an overall decrease in the percentage of persons engaged in routine work of these kinds. Correspondingly, there will be a larger percentage of employees than before in service occupations and probably also in technical occupations. The work-satisfaction studies discussed earlier show differences among occupational groups of precisely this kind. From these data it appears that if factory operatives and clerical workers decline as a fraction of the labour force, while service workers, sales personnel and professional and technical workers increase, there will be a net increase in reported job satisfaction – unless, of course, a compensating shift takes place in aspirations, a possibility we must not dismiss.

On all counts, then, we must acquit the computer technology of the charges that it has been and will be a cause of widespread alienation from work. Empirically, we find no signs of a downward trend in work satisfactions, and when we look at the actual impact of automation upon the workplace and the work force, we find no reason why such a trend should be expected. On the contrary, the newer technologies may even have a modest humanizing effect on the nature of work. The notion of a Golden Age of work prior to the Industrial Revolution must also be dismissed as romanticism, unsupported by such evidence as has been examined.

Control and privacy

The potential of computers for increasing the control of organizations or society over their members and for invading the privacy of those members has caused considerable concern. The issues are important but are too complex to be discussed in detail here. I shall therefore restrict myself to a few comments which will serve rather to illustrate this complexity than to provide definitive answers.

A first observation is that our concern here is for competitive aspects of society, the power of one individual or group relative to others.

Technologies tend to be double-edged in competitive situations, particularly when they are available to both competitors. For example, the computerization of credit information about individuals facilitates the assembly of such information from many sources, and its indefinite retention and accessibility. On the other hand, it also facilitates auditing such information to determine its sources and reliability. With appropriate legal rules of the game, an automated system can provide more reliable information than a more primitive one and can be surrounded by more effective safeguards against abuse. Some of us might prefer, for good reasons or bad, not to have our credit checked at all. But if credit checking is a function that must be performed, a strong case can be made for making it more responsible by automating it, with appropriate provision for auditing its operation.

Similarly, much has been said of the potential for embezzlement in computerized accounting systems, and cases have occurred. Embezzlement, however, was known before computers, and the computer gives auditors as well as embezzlers powerful new weapons. It is not at all clear which way the balance has been tilted.

The privacy issue has been raised most insistently with respect to the creation and maintenance of longitudinal data files that assemble information about persons from a multitude of sources. Files of this kind would be highly valuable for many kinds of economic and social research, but they are bought at too high a price if they endanger human freedom or seriously enhance the opportunities of blackmailers. While such dangers should not be ignored, it should be noted that the lack of comprehensive data files has never been the limiting barrier to the suppression of human freedom. The Watergate criminals made extensive, if unskillful, use of electronics, but no computer played a role in their conspiracy. The Nazis operated with horrifying effectiveness and thoroughness without the benefits of any kind of mechanized data processing.

Making the computer the villain in the invasion of privacy or encroachment on civil liberties simply diverts attention from the real dangers. Computer data banks can and must be given the highest degree of protection from abuse. But we must be careful, also, that we do not employ such crude methods of protection as to deprive our society of important data it needs to understand its own social processes and to analyze its problems.

Man's view of man

Perhaps the most important question of all about the computer is what it has done and will do to man's view of himself and his place in the universe. The most heated attacks on the computer are not focused on its possible economic effects, its presumed destruction of job satisfactions, or its threats to privacy and liberty, but upon the claim that it causes people to be viewed, and to view themselves, as "machines".

To get at the real issues, we must first put aside one verbal confusion. All of us are familiar with a wide variety of machines, most of which predated the computer. Consequently, the word "machine" carries with it many connotations: of rigidity, of simplicity, of repetitive behavior, and so on. If we call anything a machine, we implicitly attribute these characteristics to it. Hence, if a computer is a machine, it must behave rigidly, simply, and repetitively. It follows that computers cannot be programmed to behave like human beings.

The fallacy in the argument, of course, lies in supposing that, because we have applied the term "machine" to computers, computers must behave like older forms of machines. But the central significance of the computer derives from the fact that it falsifies these earlier connotations. It can, in fact, be programmed to behave flexibly, in complex ways, and not repetitively at all. We must either get rid of the connotations of the term, or stop calling computers "machines."

There is a more fundamental question behind the verbal one. It is essentially the question that was raised by Darwinism, and by the Copernican revolution centuries earlier. The question is whether the dignity of man, his sense of worth and self-respect depends upon his being something special and unique in the universe. As I have said elsewhere:

> The definition of man's uniqueness has always formed the kernel of his cosmological and ethical systems. With Copernicus and Galileo, he ceased to be the species located at the center of the universe, attended by sun and stars. With Darwin, he ceased to be the species created and specially endowed by God with soul and reason. With Freud, he ceased to be the species whose behavior was – potentially – governable by rational mind. As we begin to produce mechanisms that think and learn, he has ceased to be the species uniquely capable of complex, intelligent manipulation of his environment.

What the computer and the progress in artificial intelligence challenge is an ethic that rests on man's apartness from the rest of nature. An alternative ethic, of course, views man as a part of nature, governed by natural law, subject to the forces of gravity and the demands of his body. The debate about artificial intelligence and the simulation of man's thinking is, in considerable part, a confrontation of these two views of man's place in the universe. It is a new chapter in the vitalism-mechanism controversy.

Issues that are logically distinct sometimes become stuck together with the glue of emotion. Several such issues arise here:

To what extent can human behavior be simulated by computer?

In what areas of work and life should the computer be programmed to augment or replace human activities?

How far should we proceed to explore the human mind by psychological research that makes use of computer simulation?

The first of these three issues will only be settled, over the years, by the success or failure of research efforts in artificial intelligence and computer simulation. Whatever our beliefs about the ultimate limits of simulation, it is clear that the current state of the art has nowhere approached those limits.

The second question will be settled anew each year by a host of individual and public decisions based on the changing computer technology, the changing economics of computer applications, and our attention to the social consequences of those applications.

The answer to the third question depends upon our attitudes toward the myths of Pandora and Prometheus. One viewpoint is that knowledge can be dangerous – there are enough historical examples – and that the attempt to arrive at a full explanation of man's ability to think might be especially dangerous. A different point of view, closer to my own, is that knowledge is power to produce new outcomes, outcomes that were not previously attainable. To what extent these outcomes will be good or bad depends on the purposes they serve, and it is not easy, in advance, to predict the good and bad uses to which any particular technology will be put. Instead, we must look back over human history and try to assess whether, on balance, man's gradual emergence from a state of ignorance about the world and about himself has been something we should celebrate or regret. To believe that knowledge is to be preferred to ignorance is to believe that the human species is capable of progress and, on balance, has progressed over the centuries. Knowledge about the human mind can make an important contribution to that progress. It is a belief of this kind that persuades researchers in artificial intelligence that their endeavor is an important and exciting chapter in man's great intellectual adventure.

Summary

From an economic standpoint, the modern computer is simply the most recent of a long line of new technologies that increase productivity and cause a gradual shift from manufacturing to service employment. The empirical evidence provides no support for the claim sometimes made that the computer "mechanizes" and "dehumanizes" work. Perhaps the greatest significance of the computer lies in its impact on Man's view of himself. No longer accepting the geocentric view of the universe, he now begins to learn that mind, too, is a phenomenon of nature, explainable in terms of simple mechanisms. Thus the computer aids him to obey, for the first time, the ancient injunction: "Know thyself."

Information: the ultimate frontier

Lewis M. Branscomb

Not too long ago I was invited to a conference that was called to look into the future of technology. The organization that invited me, along with such distinguished scientists as Murray Gell-Mann of Caltech and Bernd Matthias of the University of California, was a government agency, and, in keeping with its policy, it did not tell us the purpose of the meeting until we arrived. Once there, however, we could understand why it was kept quiet. This agency wanted to know what big scientific discoveries were in the offing in the next two decades. When it became clear that this discussion was going nowhere, Gell-Mann volunteered the suggestion that the agency had brought in the wrong people.

"The people you need here," said Gell-Mann, "are the science fiction writers. They are skilled in telling you what will be discovered in the next 20 years. We scientists are the experts in why these visionary ideas won't work." "I agree," Matthias broke in, "but why call a meeting of science fiction writers? Just read the 30 year predictions they wrote ten years ago."

What I shall attempt here is the difficult dual task of indicating what will not happen and what will happen in information technology not only in 20 years, but in 100 years. Bravely, I shall try to do both without the aid of the science fiction writer. I will, of course, look at the technology and its limitations – which I really think cannot be done beyond 10 or 20 years – and then at the challenges posed by the powerful social changes that inevitably accompany hardware and software developments.

To place the information resource in perspective consider food and energy, which are of course basic resources of mankind: we need to know how available they are. But that is not enough. We also need the wisdom to manage their distribution. That takes information – the kind of information that does not simply end up with a count of the starving, but that provides the immediate logistics as well as the models and concepts for bringing about distribution promptly, equitably, and efficiently.

In this connection it is interesting to note that there is another striking difference between the three resources. Food and energy challenge us because they are in short supply in major areas of the world. In both cases the challenge seems to be, How do you manage shortage? Information is quite different. It is in quantitative surplus. To be sure, there are great gaps in human knowledge that have yet to be filled by research and study. But the yawning chasm is between what some people have learned, yet others have not put to use. Indeed there seems to be more information around – good and bad – than anyone can use. I have heard it said we are living in a world of information junk, which is reminiscent of critical remarks often made about certain foods.

On top of that, there are other features of information resources that distinguish them from food and energy. Information does not disintegrate when it is used: in fact, consumption generally increases its value. It need not go away – or rot – when you do not use it. And no natural law limits how much of it people can have. The more you have, the more you want, and the easier it is to get.

The trouble is, nobody knows how to measure the worth of information. It often depends on who has it, or who does not have it, and what it takes to generate information when needed. People rarely distinguish between data, information, knowledge, and wisdom. Yet they are as different from one another – and as interlocking – as starch molecules, flour, bread, and the flavorful memory of a superb morning croissant.

The question in information is, How do you handle a situation of plenty? What is even more difficult, since so many individuals are information poor, how do you use the surplus of information in society to overcome the scarcity of information available to individuals?

Information and communication: the computer

The key element of course, if you will pardon my announcing it so soon, is the computer, which is not only an information machine, but also a communications device. The two terms – information and communication – are often blurred when they are tossed about loosely, but it is important to make the distinction if one is to look at what may happen in the up-coming years. When information is available in machine-readable form it can be both processed and communicated. Processing permits meaningful manipulations of the contents of the electronic traffic, thus enhancing its value. Communicating, from an engineering point of view, means simply moving electronic traffic from one place to another. It matters little if the signal represents random noise or a Shakespearean sonnet.

Now with this in mind we can take a look at what is happening in the field of information technology. The trend of the computer industry, as we all know, has been toward computers that are smaller, faster, and cheaper, to give us more computer power per dollar. The impetus for this increase in computer power has come from several directions. The most important is the improved performance of the electronic circuitry, which has brought switching speeds into the picosecond, or trillionth of a second, range. Since the time taken for an electrical impulse to move from one circuit to the next is a major limitation on computer speed, the circuits have to be packed very close together.

Circuitry

To illustrate this point, let us say that a decade from now there is a need for a basic computer operation at the rate of ten machine cycles per nanosecond, or billionth of a second. Ten cycles per nanosecond is 50 times the speed of the fastest computer today. To build such a machine the designer will have to face the fact that an electrical signal can travel no faster than the speed of light, or only about 3 centimeters (a little more than 1 inch) in 1/10 nanosecond. To avoid deleterious transmission delays we would have to fit all the circuits into a hypothetical computer no bigger than a 1-inch cube, something you could wear on a watch chain, in order to carry out the proposed operation. (Incidentally, this 1–inch cube will be my last mention of inches. By the year 2078 I would expect this nation to be metric at last. Five years ago I helped draft the law that ordained conversion to the metric system in the United States in 10 years. I would think we might have enough time to make it by 2078.)

To get back to our Lilliputian world, I have to point out that in computer electronics – as in Disney World – small is beautiful. When integrated circuits are densely packed, as they would be in our little cube, it takes less energy to drive them fast. Furthermore, the smaller the circuits the cheaper they are to make. Such circuits – 300,000 might be needed in the hypothetical computer I have described – would certainly have to be laid out in patterns with dimensions smaller than 1 millionth of a meter, which can only be done with exotic techniques such as X-rays or electron beams. But since each circuit would have to switch at incredibly fast speeds, it would generate heat on the order of a few milliwatts. For the entire miniature computer the power to be dissipated would amount to about 1 kilowatt, enough to operate a two-slice toaster. In today's semiconductor technology – and even in future technology – it is extraordinarily difficult to draw off this amount of heat from 1 cubic inch of circuitry without damage. In short, the power laws of physics make it difficult for us to envision the development of midget supercomputers with silicon semiconductors. Silicon technology has served us well for 25 years and is likely to continue in this service for a long time to come. Yet it clearly has its limits. Other semiconductors such as gallium arsenide are under investigation. Gallium arsenide may turn out to be faster than silicon, but it too shares the limits of silicon. So we have to start thinking now about other technologies that can support computer progress in the following decades.

Fortunately, a new technology is on the horizon. It looks as if it might take over from silicon and other semiconductors for high-performance processors and give us smallness and high speed with extremely low power dissipation. Called Josephson technology after British Nobelist Brian Josephson of Cambridge University, it uses metal alloys cooled to the temperature of liquid helium, 4.2 degrees above absolute zero. At this temperature, where virtually all molecular motion ceases, many metals lose their normal resistance to the flow of electricity and become superconducting. When two such superconducting metals are put very close together in the form of a thin sandwich, a magnetic field from a nearby current source determines whether electrons will bridge the gap. The gap, by the way, actually contains a dielectric a dozen atoms thick. The electrons set up a rivulet of electricity that tunnels through the sandwich and switches the device.

In IBM's laboratories at Zurich and at Yorktown Heights, New York, fragments of computer circuitry have been built on the Josephson principle and have operated successfully. Individual circuit devices have demonstrated a switching time faster than 20 trillionths of a second. There are still difficulties to be overcome with this technology, but it promises to lead us to a computer that is very small, very fast, and very cold – and, I might add, very expensive, at least initially. In 1973, IBM's Leo Esaki and General Electric's Ivar Giaever shared the Nobel Prize for Physics with Brian Josephson. The three were honored for laying some of the scientific foundations of this new technology of superconducting electronics. It is important to note that these scientists, along with another trio, John Bardeen, Leon Cooper, and Robert Schrieffer, who shared the prize the previous year, have made it possible to understand the fundamentals of Josephson technology at least as clearly as silicon technology. Because the theory is so well established, we are hopeful that progress in Josephson development will be rapid.

Memory

Let us now look at the memory side of the computer and see how it too has changed and where it might be going in its own evolution. One way of approaching the development of computer memory is to compare it with another great intellectual storage facility, the library. The information content of a huge library such as the Library of Congress can be reduced to the language of bits by simply doing a little arithmetic. We are told that, as of September 1977, the Library of Congress had 18,320,256 books and pamphlets. Let us give them credit for 20 million books today. If there are 300 pages in a book, with each page consisting of roughly 1500 letters, this gives us 450,000 letters per book. The 20 million books would thus become 9×10^{12} letters or about 70 trillion bits. Now the IBM 3850 Mass Storage System will hold 3.8×10^{12} bits, which by a process of division informs us that the volume of books in the Library of Congress could be stored in less than 20 IBM 3850's. So far I have been talking about a future that is almost here, not about the year 2078. To stretch that far I am going to have to rely on the tools of the technological forecaster, since I am not a science fiction writer. But I should warn you that I am on record as ardently opposed to the use of graphical extrapolation as a way to predict the technical future. In fact, I have publicly advocated a ban on the sale of the main tool of the forecaster: semilog graph paper. Since nobody has taken up my proposal, I will now use semilog graph paper to examine the rate of growth of computer storage capacity available at a given cost – which has averaged nearly 35 per cent a year for a couple of decades. Even if that slowed down to 21 per cent a year in the future, we can see that in 100 years the same investment that would hold 20 million books in computer storage today would finance 1.5×10^{10} or 15 billion electronic libraries, one each for the 15 billion human beings I expect will be living on the earth in 2078.

Of course, this is a rather daring extrapolation of progress in magnetic storage, as well as in a few other fields. In magnetic storage it implies a surface area of only 3 square angstroms, or the area of one atom, for the magnetization that stores one bit. That is a very small magnetic domain, and I am not sure that this prediction will come about. Library storage, in terms of information per square centimeter, is fairly primitive when you consider the storage potential of the human brain, and here the figures that we can project become more audacious, if that is possible. Let us, for instance, quantify the capacity of the brain in terms of bits by assuming that each synapse is the equivalent of a storage element. On the best physiological evidence the brain has 10^{13} synapses, or a comparable number of bits. Twenty five years ago a computer memory packing this much information would have filled a small mountain 500 meters high. Since 1953, main computer memory has shrunk 800 times in size and is continuing to shrink at the same rate. If we again degrade this rate of progress to 21 per cent per year and project a century ahead (more semilog graph paper, please), we come to the astounding conclusion that the information density of the computer will actually outstrip that of the brain – assuming that the pace of human evolution will not suddenly speed up. Of course, the comparison is not exactly fair. There are tricks of association with which the brain stores information that somehow the computer has never matched.

Nonetheless, on the basis of the growth indicated, the computer of the year 2078 will contain the data memory equivalent of 16,000 human brains – the equivalent of a university population, including graduate students. While I am at it, I might add another improbable figure to this project. That figure is cost. If today's largest computer costs $5 million, and the cost per bit of on-line memory and per circuit of logic is decreasing at the rate of only 21 per cent a year, then in 2078 the cost of a computer with the power of today's big machines will be extraordinarily low – in fact, 15 billion times lower, or down to 3/100 of a cent. Nobody believes, of course, that this will happen. I expect that some of the largest computers of the year 2078 will have to sell for around the same price as today; of course, the power and usefulness will improve enormously – perhaps by that same factor of 15 billion. All in all, it does seem unlikely that semiconductor memory – or "magnetic bubble" memory, a new form of storage that may replace magnetic disks – will continue to shrink in size and cost for an other 100 years at the same rate we have enjoyed in the last decade. In fact, if we are still using semiconductors 100 years from now and have invented nothing new, we would have only three atoms of space to work with in the memory cells of the computers of that era. This is not much, so we will have to invent something else.

As far as I can see, that something else will have to be a complex three-dimensional memory cell, which will have the power to self-replicate. Do we have a model of this? Of course – it is a strand of DNA, which has a genetic memory of 10 bits of information and is programmed to rebuild itself from its own chemical storage to accommodate new experiences. A molecule of DNA not only has a specific chemical composition, it has a specific structure unique to its information content. Before the 100 years are up, we will probably have to build a biological crystal computer out of something like DNA; in short, we may have to reinvent the brain. Science fiction, of course, has arranged that for us already, which shows that Professors Gell-Mann and Matthias were on the right track. Perhaps we could have found the answers to the computer future in the science fiction plots of 100 years ago.

Access to information: the next 100 years

So far in this survey we have been talking about information already inside the computer, where it moves at high speed and inexpensively. It is interesting to note, however, that the hardest problem concerns how information gets into the computer in the first place – and how you get the needed results out.

Here, too, new technology is on the way to help. The computer of the future will not have to be approached by push button or typewriter key or a punched card. It will respond to speech. This capability will take a decade or so to be fully realized. Then you will need only tell the computer what it should know to have the computer respond. Computers of the future will also be able to "read" by pattern recognition more rapidly than ever, so that visual processing will also become natural to the new and advanced generation of computers. The tyranny of the typewriter's limited character set has already been broken by computer-processed pictures and symbols, which will be added to text to enrich our "written" languages. The computer's ability to read symbols by using pattern recognition logic will be matched by its ability to write symbols, using display screens and nonimpact printers. Two emerging examples of nonimpact printer technologies are printing by ink jets and printing by laser electrophotography.

This trend, already well under way, will ultimately completely obscure the technical distinctions between information technologies. Typewriters, television, movies, telephones, and even radio, records, and tapes will all become interrelated and interchangeable. The printing press is going electronic too, but despite the electronic substitutes the one assured survivor among 1978 technologies in 2078 will be the lowly piece of paper, which will doubtless be made of synthetic cellulose. Many paper jobs may be taken up by electronics, but some uses will be left. Certainly, scientists will still work out original calculations on the backs of envelopes or on restaurant napkins. The letter post service as we know it will have gone out of business by then. People will have to manufacture paper envelopes just to carry them around in their pockets.

In the next 100 years, much information that is stored today may not have to be stored at all. It may be cheaper to reconstitute or reacquire information from basic elements each time the information is needed. It is certainly easier to rebuild a library after it has burned down if its books are stored electronically in safe remote locations.

Today mathematical tables are obsolete. Scientists in the future will doubtless be startled to discover that the 50th edition of the *Handbook of Physics and Chemistry* on my bookshelf actually contains a table of the reciprocals of the integers. Why print such a table? By storing a simple algorithm you can order the computer to build the interesting parts of a table whenever you want them. In some cases the database is inherent in the world around us. It may be easier – and better – to reobserve something than to retrieve a stored recollection of it. For example, it may be easier and better to ask a satellite to reacquire an up-to-date street map of Chicago when you need it than to rely on city officials to provide current records. We can also obtain updated weather observations, and even actual commuter train arrival times based on direct observation of the train rather than on timetables.

In our present 20th-century world of print and paper, we tend to think of information in terms of documents. In the future, our information machines will permit us to enjoy more immediate access to all kinds of information-gathering capabilities. Documents will become only occasional by-products of information access, not the primary embodiment of it. I can, for instance, see a television viewer of the future ask his home computer for a selective printout of the day's news after he has already heard the headlines and seen some of the action displayed on the screen. He can then remove the written newspaper prepared by his computer and read it at his leisure.

Information interchange will be furthered by advances in communications technology. Well before the year 2078, for instance, the broadest bandwidth capability – for data, voice, and images – will be widely available by means of optical techniques using glass fibers and lasers. A single glass fiber the diameter of a human hair can carry 800 voice conversations, tens of thousands of data messages, or 50 million bits per second – in short, enough capacity to carry the contents of more than 40,000 typical books from the Library of Congress in Washington, D.C., to Los Angeles in 1 hour. At very little additional cost, hundreds of fibers can be packaged in a single cable. These techniques will reduce enormously the cost of communications. Optical technology can eventually be expected to take full advantage of the bandwidth, or traffic-carrying capacity, of the visible light laser – about 600 trillion hertz, or 100 million times the capacity of today's optical fiber systems. That is 1 billion books per second on each optical channel, instead of 40,000 per hour. Even terminals may be affected by light wave communication. IBM is experimenting with a novel approach that permits terminals and hand-held devices to communicate with other devices in a room without using wires. Each device sends out and receives invisible infrared light pulses that are reflected off the walls. In the second half of the 100-year period, incidentally, space satellites will no longer be used for point-to-point communication on the earth. Instead, they will be reserved for collecting data from remote or mobile automatic stations, or for broadcasting information simultaneously to very large numbers of receiving stations. No doubt they will also be used to engage in what Anne Branscomb calls "the ultimate diplomacy" – receiving and interpreting information from sources of intelligence in outer space – and beginning what is likely to be a very protracted dialogue.

Personalized manufacturing and service

The technology I have discussed so far does offer some challenges for the future, but I suspect that future generations of engineers will be able to meet them successfully. There is another kind of promise that can also be met and that is the promise of personalized manufacturing and service. This is a great step forward from the Industrial Revolution, where economy of production depended on the sameness of each item produced and the manufacturer usually passed service responsibility on to someone else.

The computer, in my view, will make it possible for people to overcome the deficiencies of the Industrial Revolution and at the same time belie the Marxian assumption of worker regimentation arising from mass production. The fact is that with the computer, manufacturers may for the first time make articles on a mass basis, yet have each fit the specific requirements of the intended end user. This is, in fact, the way IBM manufactures custom logic circuits under computer control today. Many other examples can be envisioned.

Let us take a futuristic example from the shoe industry. People are forced to fit their feet into a few basic shapes and sizes of shoes. As a result of imperfect fit, foot trouble is one of the most common medical complaints. But the last on which the shoe is made can be infinitely variable if the shoe machinery is computer-controlled. The computer can take detailed measurements of a person's foot and communicate the data along with style preference to the factory. Then an individually fitted shoe can be made at mass production prices.

All this involves a new personal relationship with the computer, but humans, being intellectually inefficient creatures, may be unprepared for this new interactive role. In the future, the computer will be able to assist people in overcoming their own weaknesses in communication in many ways. It will do this by providing access to incredibly wide worlds of information, along with instructions on how to best use this information in a particular circumstance. This will require development of high-level computer languages, more like natural English. With the help of such a capability we will only have to ask the computer what help it can provide with a particular problem and the computer will guide us through the alternatives to provide an authoritative individualized approach. This may make the computer seem all-powerful – which some mistakenly believe is true today – but I think it is important to distinguish between a powerful servant, which the computer is, and a powerful god, which the computer, for all our hyperbole, is not.

Communicating through computers: human factors research

To reach this stage of service, computer software will have to be designed with people in mind, not to replace them, but to let people relate to machines in ways that more nearly resemble the ways people relate to one another. Today, two high-speed computers talking to one another through narrowband telephone channels are like two lovers trying to communicate through the post office. But even when communications bandwidths match the internal speeds of computers, people will realize that there is more to communication than the exchange of packets of data.

Consider sociability, the warming-up process that precedes good communication. My wife was involved recently in a computer teleconference project with a number of academic people in different states. Each participant had a computer terminal in his or her home. One professor could not relate to the system. He was impatient with the way his colleagues communicated with one another through the computer. "When I log on to the computer," he typed, "I see personal messages such as, 'how are the kids,' and so on. I don't like to waste time. So next time you want to use the computer, why not just type 'chitchat' before you send the serious message. I'll know you mean to be sociable."

The interesting thing about this reaction, my wife reported, was that when the professor uses the telephone for business, he always starts the conversation with some social comments about friends and family. She suggested that he simply say "chitchat" and get on with it. He disagreed. Evidently he felt the need to start with informal comment to set the mood on the communication instrument familiar to him. But the computer was so forbidding and unfamiliar that he could not relate to it.

Research on this problem – making the computer more friendly to its human masters – is known as human factors research in the United States and as ergonomics internationally. Alphonse Chapanis of Johns Hopkins, who is president of the International Ergonomics Association, did a pioneering study of how people communicate using machines. Observing that most computer languages are highly structured, with disciplined syntax, he wondered if this reflects the way people prefer to communicate when they need to exchange information. To find out, he divided his laboratory into two rooms, each with a terminal and a person. In one room he placed on the floor pieces of a do-it-yourself pushcart. In the other room he provided a set of assembly instructions. He observed the time taken by the individuals in the two rooms to assemble the cart through communication by typewriter terminal. Several pairs of individuals were tested. He found that the job was done most rapidly when the people sent each other fragments of seemingly inefficient verbal traffic. They rarely communicated in structured sentences the way programmers do.

So we may want the computer of the future to generate some sociable chitchat to test the mood of its human master before communication begins. And we may want the computer to generate spontaneous, fragmentary communication, thus making it just human and inefficient enough to be truly effective.

The integration of the computer into the communication patterns of the future will continue to have a powerful impact on many of our social and political institutions. In education, for instance, computers are currently useful but limited tools, yet they hold the potential of altering the role of educational institutions. Let us take universities for example. Today universities are not just generators of knowledge but also distributors, making them poorly paid partners of publishing. Electronic media – television in our time and interactive home video computers in the future – will change that by providing even more effective mechanisms for education – indeed, more effective than the classroom because of the personalized nature of the educational processing. Marshall McLuhan understood this trend years ago when he said, "Children resent time spent in school because it interferes with their education."

In the future, with electronic systems to distribute knowledge, the university will be free to give greater attention to the generation of knowledge. This was its role a century ago – being the center of scholarship. New institutional arrangements that stress human relationships between student and faculty will replace the rigid structure of schools and classrooms we have today – to achieve universal, lifelong education productively, effectively, and in response to each person's needs. For this to happen knowledge will have to achieve its proper value, and knowledge creators, in universities and elsewhere, will have to be paid accordingly by the consumers of knowledge for full value.

Today the worth of this knowledge seems to depend on how well one's text book sells. This, in turn, depends on marketing and distribution. If we can solve the problem of protection of intellectual property, by the year 2078 there will be an effective market for knowledge in itself, since distribution to the consumer will be easy and direct and very inexpensive.

The computer-communications systems of the future will also continue to stimulate the further development of our liberties and the institutions that protect them. The First Amendment to the Constitution, for instance, concerns the freedom of speech and press and was designed for an era when liberty was protected by the right to buy a handpress and distribute one's own leaflets. Today this is not enough. Very few of us can buy a television station or get a license to operate it to guarantee our right to make a speech.

I believe that we must expand the citizen's right to acquire information and to express himself through the existing media system, as well as to be protected against the misuse of information about him. With the advance of widespread access to the media, which people will have to be assured of in the future, I think that information technology will be able to help guarantee these fundamental rights of equity and privacy and to redefine them for inclusion in the codes of our future society.

The computer has often been accused of being the bureaucrat's tool – the prime instrument that enables government officials to limit our freedom of movement and manipulate information against our interests. Solzhenitsyn, in *Cancer Ward,* makes information bureaucracy very graphic when he writes that "as every man goes through life he fills in a number of forms for the record. ... There are hundreds of little threads radiating from every man, millions of threads in all. If all these threads were suddenly to be come visible the whole sky would look like a spider's web and if they materialized as rubber bands, buses, trams and people would lose the ability to move." Solzhenitsyn went on to express concern that such a cobweb could be manipulated by those in a position to control the threads.

For this Russian author the illuminated cobweb symbolizes strangulation and suffocation by bureaucracy in an authoritarian society. But I think a cobweb is also a description of the social interactions that give us the basis for a democratic society. Without informational interlinkages of this kind we could not have developed the infrastructure that guards our liberty and security, while we seek to share the world's opportunities and resources. Of course, some societies may allow the infrastructure to evolve in a disorganized way – like the mess of ticker tape that inundates a visiting hero in New York – and we may end up with a very confused world. Or one can deliberately unravel the fabric of society as advocated by those who reject contemporary society and seek the simple life in isolation.

Fortunately, there is a better alternative. We can also learn to manage the cobweb, making subwebs, building proper couplings, leaving lines and linkages intentionally vague in some cases and making them precise in others, and so arranging the overall pattern that society maximizes the return to the individual from each interconnection. The principal challenge of the future will be to make this pattern of interdependence ever more flexible, practical, and rewarding for all of us. This is the true challenge for information systems in the year 2078.

The informatisation of society

Alain Minc

The report on the informatisation of society is a hybrid document; it is in fact based on a technical assessment which I shall little expand upon since you are infinitely more expert in these matters than I am. This is the combining of telecommunications and automatic data processing, for which a French barbarism, "télématique", has been coined. What we discovered was that the combination was an important one and that it coincided with a whole series of parallel revolutionary events affecting computers and telecommunications: in regard to computer technology this is the appearance of microprocessors and the emergence and forthcoming development of an inexpensive, easy-to-handle and extraordinarily widespread type of data processing. This, at a time when telecommunications are themselves evolving, in fact means that the traditional divisions between television, telecommunications, radio and data transmission are being broken down.

Clearly the climax will be the advent of satellites of a composite type which will equally well transmit vocal messages, pictures and sounds and provide a fantastic instrument of communication. This is the gist of the technical assessment, which I shall develop no further, but which in the eyes of the public represents a relative breakthrough, since it has regarded the computer as a cumbersome, unmanageable and discouraging tool while telecommunications have been relegated to the telephone line alone.

From this assessment three types of consequence emerge, which concern:

1) a problem of sovereignty;
2) an economic analysis relating to the productivity gains yielded by such a tool;
3) questions as to the long-term development of society.

Sovereignty comes into play at two levels:
1) control of networks,
2) control of data banks.

I shall deal fairly briefly with the problem of networks, which is that the natural growth of such transmissions will largely be governed by the marketing strategies of various firms and to a considerable extent by the leading world computer enterprise. This is quite normal and legitimate, but it is clear that among the various manufacturers and network managers separate territories will gradually be built up. There will be an

IBM territory, in which the different networks set up by this firm will be interconnected, and which will adjoin the territories of other manufacturers. From this standpoint the danger is that virtual frontiers will be traced between territories of the various manufacturers: communications from one to another will therefore be impossible. To take another example, let us say two adjoining towers in the Défense district: if one of the towers belongs to some manufacturer's territory it will have no trouble in communicating with any enterprise in the same territory as far away as Tokyo or New York, but will be unable to communicate with some neighbour 200 or 300 metres away belonging to another territory.

What we are trying to say is that States have so organised themselves as to provide for the transparency of communications, and that a challenge they have to face is the possible emergence of teleprocessing and electronic frontiers. Their task is to regain control of transparency by intensive standardisation to make intercommunication possible. In speaking of States, we do not mean to imply any nationalistic, chauvinistic approach; to our mind none other than an international policy of standardisation can be considered. We would prefer no standards at all, hence electronic frontiers, rather than standards which might be purely national. What we therefore recommend is that our Government be the vehicle of a standardisation policy such as that developed by the CCITT for land-based networks with the X25 standard.

For land-based networks it is necessary to go further, and in future also to try and have the same type of international standards universally accepted for satellites. In our view this is an important line of action, since it involves a number of proposals for internal re-organisation bound to affect our administrative world, the most difficult and painful step in France being to separate telecommunications from postal services. We thus consider that the type of international strategy here required cannot be conducted by an administration which in its day-to-day management is subject to the constraints of co-existence and co-habitation with the postal service, which is of an entirely different nature and with which in future years it will be competing directly.

The second aspect of sovereignty, which I merely propose briefly to mention here, is the development of data banks. This, I think, is perhaps a more acute problem in France, since our policy in this regard is less advanced than in other countries – the United States, of course, but also other countries here represented. We believe that a distinction should be made between two types of data bank – the fact that data banks should not be nationally available in such highly technical fields as chemistry, materials resistance or any other technology we do not regard as a fundamental problem so long as access to foreign data banks is easy, satisfactorily priced and does not threaten to be cut off. Hence on this score we have no special comment, unless it is to want connection with existing banks.

But what we do claim is that there are non-technical fields generally having to do, say, with the human sciences where data banks are being built up, mainly in the United States, and that in this regard the possession of data banks is a problem of an altogether different kind, since that of a cultural model is involved. I shall mention one example which we have tested, as it were, and which we think to be very meaningful. Today most French newspapers are being connected to the best existing news data bank, which is that of the New York Times. Thus if this bank is questioned regarding certain episodes of French political history, a very interesting picture is obtained, one which of course very naturally depends on the views which reporters of the New York Times may have held but which is not exactly the same as ours concerning our own history. We do not claim that the latter should be the only view, but there should at least be two ways of looking at our history. We already have one interpretation, that of the New York Times; we should now arrange for also providing a French viewpoint, since within the next ten years any journalist of a local, peripheral newspaper connected only to the New York Times would obtain a picture of French political history which, while interesting, deserves to be compared with some other slant.

A second [example] has to do with history proper, where clearly the way information is structured is altogether fundamental, and at research level is responsible for many of the conclusions which are drawn. In our opinion this means that here again data banks should provide a dual picture in relation to our own model of cultural development. While this statement may strike you as extremely nationalistic and very traditionally French, I do want to point out that I hold no such brief concerning any discipline of a technical nature, where I believe intercommunication goes without saying and where no nationalistic problem arises. But I do so argue, very definitely, in the case of anything which seems closely to affect our cultural model and cultural identity. A sort of hard national core is involved here, one which we all have and which, unless we are careful, data banks can cause either to shift or to change.

The second subject for thought is of an economic nature. The evidence clearly shows that in forthcoming years data processing or teleprocessing, taken in the broad sense, will lie at the root of all productivity gains. These are either achieved by some direct investment, or else this investment enables all other investments to be made. To day, particularly in industry, no investment for purposes of modernisation fails to make use of these techniques in some way. In coming years there will be a manifest opportunity for improving productivity, one all the more significant as a number of traditional services which so far have been counterproductive can be brought up to a satisfactory standard. A problem of a two-fold nature here arises:

- in the first stage an employment slack factor which must be recognised;
- in the medium term an employment recovery factor.

It is clear that in the service sectors where networks are set up and where all the relevant internal re-organisation is carried out, productivity gains directly affecting employment in the undertakings concerned will occur. By this we do not mean dismissal but non recruitment, which in terms of recourse to the labour market are quite similar.

The main sectors concerned are:

- large service purveyors, such as banks and insurance companies;
- such an all-pervading area of the economy as office work, particularly secretarial tasks;
- the postal sector, where due to the sharp competition of these new techniques, postal activities are bound to decrease in the long term;
- the internal services of major industrial firms. Thus in enterprises now employing 5,000 people, for example, the share of indirect services often accounts for over half of the staff employed.

These employment slack factors complicate an already difficult situation. But once the Malthusian reflex – a meaningless reaction – is eliminated, since in any event competition will impose measures of rationalisation, such techniques even so would seem to offer job opportunities. For purposes of argument let us revert to such an altogether fundamental factor and major economic constraint which the balance of foreign trade has become. Clearly the deteriorating employment situation in France today calls for one basic aim, which is to close the trade gap. To succeed in this we are running with all the brakes on: limiting the money supply, limiting the budget deficit and limiting transfer policy in order to meet a priority requirement of any economic policy, which is to bring foreign trade back into balance. And it so happens that another side of the productivity gains I spoke of is that if intelligently channelled towards foreign trade, we can better and more easily leap the hurdle of competitiveness.

I admit that here I am eluding a fundamental issue, which is how productivity gains scattered throughout the economy can be channelled towards the export sectors so as to overcome the competitiveness obstacle, but I shall refrain from labouring the point since the type of debate and problem involved will readily be realised. What I would however like to emphasise is use of the economic freedom regained by closing the trade gap regarding the creation of new jobs. We do not think that closing the gap will alone be enough to promote full employment. The reason is that, in view of the type of specialisation which each country here represented has to face, we clearly will often have to lay off more people in order to become competitive than we can take on again by increasing our outlets on foreign markets. On the other hand, after the gap has been closed the brakes on all measures will be released so as to make the new demands economically feasible.

By new demands I have in mind especially those in the sector of community services. In our country and I imagine in most of yours, there is still a considerable latent demand in health care, education, social services and transport. But there is also one for other kinds of service which has yet to make itself felt. What actually prevents these services from creating many employment opportunities is that they have not been made economically feasible, and if they have not it is because we have to redress our foreign trade balance. Therefore here, we think, is a sizeable reservoir for jobs of a different kind which will find expression on markets and also by transfer as soon as the bottlenecks of economic policy and the external constraint are removed.

This being granted, in speaking of new services which are seedbeds of demand I mean those traditional services we are all familiar with and see about us, such as health care, education, transport and so on, as well as other services which I have deliberately referred to elliptically. We in fact feel that there are a whole series of services whose nature we do not altogether understand, as in community or group activities, or in what go by the name of amenities, which answer a particular need. But it would be inappropriate to define them from the centre, which in our case is Paris. These can only be set up and made economically feasible by decentralised, local authorities. Hence this goes hand in hand with a change in the division of authority.

After this report had been written a fact which struck us, in happening to come across a United States statistic which seems to be quite significant evidence, is that the ratio of the States' budgets to the Federal budget, which in 1950 was some 60 per cent, has now risen to 180 per cent. This went together with a trend of job creation through initiatives increasingly taken at non-federal level. I shall omit an aspect which would lead us too far astray but deserves some thought, which is that traditional Keynesianism, where the central authority is in control, is on its way out. Without wanting to and without necessarily realising it, we are entering a kind of Keynesian era with many transactors. A number of public and economic transactors may thus be seen to be emerging, each with their own method of action, but whose juxtaposition in terms of general economic cohesion will cause various problems to arise in the next few years.

Based on this economic analysis, the next subject I propose to deal with is what we have merely claimed to be guesstimates regarding the long-term future of industrial societies. On this score I should like to mention three types of problem. The first has to do with the conflicts which will continue in these societies, the second with the capacity of different types of social systems and organisation to exist side by side in them, and the third the direct effects of teleprocessing on such vehicles of culture as language and learning. The conflicts we have witnessed and all the ideological debates between liberalism and Marxism that have revolved around them took place in the world of production, mainly that of industrial production.

In making the transition to a system which some have termed the "information society" (I am thinking of Mr Porat's work quoting a figure of over 50 per cent of the labour force in the information sector), with all the ambiguities inherent in the term, we are coming to one where the numerical, quantitative aspects – hence the conflictual, ideological consequences in the world of industrial production – will change. It is clear that in the long term the core of social conflict will not lie in industrial production. Two hypotheses then follow. One is that we are entering an era of calm devoid of conflict, rather a dreamlike concept of post-industrial society. The other is that we still remain in a conflictual type of society but where the conflicts are many, varied and decentralised, and where the stakes are entirely different. It may be in regard to housing or health, or education, social life, community management, or the firm, or anything else. The sociopolitical consequences of such a situation cannot be ignored: this means that there no longer is one unifying core of conflict.

Once we have a society where scattered types of conflict occur and where people band together according to some particular kind – thus an individual may be in conflict with the production system, health care, urban development or environment policies – clearly the modes of aggregation of various types of opposition, which lie at the very root of political action, are bound to change. In the long term this will have marked consequences for the balance of sociopolitical forces. To be sure, the change will be less conspicuous in countries where industrial conflicts have less been an essential part of community life than in countries with such a rigid ideological structure as France has. But the effects will be significant nonetheless.

A second line of enquiry is the following. Upon examining the long-term effects of informatisation, it will be noted that an industrial sector turned towards the outside and responsible for the balance of foreign trade was made to co-exist with a sector inwardly directed towards these new jobs, amenities and activities. Actually the position is much less clear-cut and the divisions are less marked. What are we now likely to see existing side by side? Very large industrial units in sectors where the law of increasing returns will inevitably continue to exert its effects, since there will always be sectors with huge units. "Small" is not invariably "beautiful", or to be more accurate a "small" approach is not always possible, in spite of what Schumacher has said.

Next to all this there will be a very broad sector of small and medium sized industrial enterprises, hence strongly influenced by the market and unavoidably by the market's inherent constraints. And behind this sector will lie that of amenities and new jobs, where the concepts of productivity and efficiency are entirely different. In fact we are already living in such a type of differentiated society, which will merely tend to grow.

Between these three areas is a whole set of gradations. Here a quite fundamental question arises for which there is no answer and which is how a society can accommodate such varied modes of social organisation. Will not this result in extraordinarily violent tensions, conflicts or hardenings of position? In fact this is tantamount to saying that in the long term only a society with a split personality is possible, that is capable of being highly productive on the one hand while easy-going and convivial on the other, the two sides being inexorably linked together yet subject to tensions which warrant some considerable thought.

The third line of enquiry I shall but briefly allude to, which is that as computerisation and teleprocessing techniques become generally widespread, a permanent, much broader relationship will result between the individual and the machine. The language used to converse with the machine will hence be much more widely disseminated. But as we all know, however refined it becomes and whatever efforts are made to make it like ordinary language, it will remain of a very special nature. Since it will spread throughout society and virtually take its place beside our ordinary, richly semantic language, the problem is who will continue to develop, maintain or hold on to the latter? Will it not become poorer?

Related to this question, as we all know, is a problem of social organisation and in fact social differentiation: language is above all a tool, an unconscious means of social differentiation. Are we not heading for a split between an intellectual elite which will be the virtual masters of traditional language and a society which will see its own customary idiom change owing to the irruption of one poorer by nature, the language used for dialoguing with the machine? I do no more than state the problem, but again a number of questions obviously arise in terms of social reorganisation and the influence exerted by various social groups. Since I lay no claim to forecasting long-term events and do not believe in futurology I shall go no further, beyond saying that I do not think such phenomena lie in the future and instead that the seeds – and often more than the seeds – are here already. The problem is that the transition from seed to forest raises questions which concern all of us.

A new concept of information society

Manfred Kochen

Introduction

"Information society" is a much overworked expression that has been used to denote many different concepts. The general sense in which it is most commonly understood refers to a growing, high-technology, materially affluent service society in which information rather than raw materials or energy is the dominant technology (Bell, 1973). This characterizes one of two possible directions in which post-industrial societies (also an overworked term) may be heading. Another form that post-industrialism may take is that of a society that reverses the trend toward complex, depersonalized systems. It strives, instead, toward a society characterized by increasing responsiveness to individual human needs, and toward preserving ecological balances. This dichotomy was stressed by Marien.

The stress on large information systems in the generalized Bell concept suggests that the conscious, cognitive, rational components of our lives are increasingly predominant. This is the most widely accepted concept of "information society" at present. The "Small is beautiful" concept with its stress on self-reliance and ecological balance reflects a yearning to return to nature, to "restore" the experience of harmony. Many other concepts or visions of society have been advanced that are at least as plausible as these two. One stresses the quest for power (Realpolitik), leading to an increasing fraction of societal resources being allocated to building incredibly complex weapons systems, resulting in an overarmed and insecure world society. Yet another stresses emotional (affective) and social forces that may be more, rather than less, important than intellectual-cognitive forces in modern, highly automated societies. Yet another concept is that of a society that stresses conative (relating to free will) forces, and yet another stresses spiritual values.

The common predilection to dichotomize has led Marien (1981) to blur these distinctions and to identify many of the features he dislikes with the Bell concept, calling it either "an information society" or, equivalently, "post-industrialism as a service society." Altogether he assigns it the following attributes:

1. Shift toward sensate, hedonistic, pragmatic rather than spiritual culture.
2. Stress on objective scientific understanding and technological know-how, particularly reductionistic specialization rather than holistic approaches.

3. Stress on professional expertise of meritocratic elites rather than emphasis on self-help and participation in decision making and organizational redesign.
4. Increasing military capabilities vs. development of global peace-keeping and arms control.
5. Stress on material affluence, as opposed to broader quality-of-life measures.
6. Urbanization.
7. Stress on advanced quaternary occupations rather than on self-service, job-sharing, self-sufficiency.
8. Social engineering and professional planning meshing with established institutions on a large scale instead of rethinking "progress" based on human needs.
9. Economic freedom and reindustrialization rather than decentralized ecologically-oriented small communities.
10. Extension of modernity toward increasing complexity and large scale institutions, with continuing fragmentation, rather than stress on synthesis and adaptation to human scales and values.

Herman Kahn (1982) has called "The Great Transition" the 400-year period that we are now in the middle of, in which most of the world's people are shifting from few to many, poor to rich, primitive to technological, rural to urban, sacred to secular, royal to administrative, parochial to global, spears to nuclear. This has been judged to be the only 400-year period (1776–2176) between 8000 B.C. and 8000 A.D. in which the population growth rate exceeds 0.1 per cent per year. Kahn regards as still viable the concept of an ever-expanding economic pie for Western societies and the "revolution of rising expectations" for the Third World, and views the "limits to growth" and "Small is beautiful" view as a Western malaise. While aware of limitations in "The Great Transition", he states it can be self-fulfilling and "influential only if it is disseminated, taught, understood and believed."

Despite differences on the issues of growth, affluence, and technology, most futurists appear to agree that (1) we are in the midst of a great transition, that "this decade is probably the most important in the history of the human race" and (2) that values, represented in symbols and language, are "the dominant, driving variables in all economic and technological systems". They agree with the slogan, "Thinking Globally, Acting Locally", to further the preservation of cultural variety and diversity. Strong concludes from this that "information networks have to be broad, complex and even loose. Action centres, however, must be small, tight and therefore mobile, responsive, flexible." He contrasts information for decision-making with action for deciding.

The simple concept of an information society as one in which information rather than materials account for most of the exchanges throughout societal communication networks is common to all these diverse visions of post-industrialism.

But why the concern with the concept of "information society"? How will it be used? Why is a new concept needed? How is it to be evaluated?

First, there are those in a society who believe its members or leaders can shape the kind of society they would like it to become. If such persons have a sufficiently clear vision of where to aim and lead and if they are actually able to realize these goals, then a sharper and deeper conceptualization of the alternative future shapes of what their society could realistically evolve into is important. Such conceptualizations are prerequisite to the recognition of options and to wisdom in choice. As suggested above, the alternative shapes are at present far from clear and distinct.

It does appear to be a next stage in evolutionary history. Brains and nervous systems in individuals have evolved to enable man to shift toward more sophisticated information-processing to control his life and environment. Analogously, new social organs may be evolving that not only serve communities the way brains serve individuals, but that go as much beyond the highest functions an individual mind performs as the individual mind goes beyond the mind of a reptile. A community brain supplants an individual's brain by exhibiting functions at levels that are an order of magnitude above those of an individual's comprehension or inventiveness. Scientific progress resulting from scientific communities is an example. Perhaps cooperative modes of political, economic and social interaction may follow this example in mature information societies.

A new concept of information society is useful if it makes more appropriate distinctions than the existing ones, so that those who select and attain goals are able to justify their choices more realistically and persuasively. It should promote greater awareness and perspectives about the major social choices. Only if this brings about societal consensus is there any possibility of major societal change.

Pre-industrial "societies", such as India, seek urgently to eliminate the back-breaking drudgery involved in much of the manual work that predominates by means of mechanization and upgrading the skills of the work force. Some leaders in such less-developed countries (LDC's) are questioning whether they should steer the course of their development by what they perceive to be the example of Western post-industrial societies, in which the service sector, and in it particularly the knowledge industries, predominate. The question should perhaps be asked to what extent the course of a society can be steered at all. A few, who are committed to such a course, would even like to omit the stage of becoming first an "industrial society" and skip directly to the information/service version of a post-industrial society. This is not likely to happen, for the following reasons.

In India, for example, 41 per cent of the gross domestic product is in the primary sector: agriculture, fishing, mining, quarrying. This is growing at 3 per cent per year. It is characteristic of a pre-industrial society. Manufacturing, construction and electricity accounted for 23 per cent. It grew at 7 per cent per year in 1981, but its percentage of world industrial output declined from 1.21 in 1963 to 1.19 in 1977. (Compare with 0.46 per cent for Taiwan in 1977.) Services, primarily financial, transportation, communications, trade, public administration and defense comprised the remaining 36 per cent, also growing at 7 per cent annually.

The majority of India's 670 millions is still rural. Its future, like its past, probably lies in its 576,000 villages, where 80 per cent of its population lives, despite the projections that by 2000 the present rate of urbanization (4 per cent per year) will result in five megalopolises (Bombay-Ahmedabad-Poona, Calcutta, Madras-Bangalore, Delhi, Lucknow-Kanpur) with populations of 38.6, 23, 19.7, 19 and 10 million, respectively. The drudgery in the primary sector activities may thus shift to the manufacturing and the service sectors. The drudgery of the illiterate peasant or factory worker is perhaps perceived with much less disdain than that of the literate white-collar worker who is frustrated by the underutilization of his education. The large number of unemployed or underemployed intellectuals engaged in work they perceive as drudgery may be by far the major source of discontent. (Much of the above probably has its counterpart for Egyptian society, which has for millenia been closely tied to the valley of the Nile.)

It is instructive to examine why the 10 newly industrialized countries (NIC's – e.g. Singapore, Hongkong, S. Korea, Taiwan, etc.) attained a greater rise in industrial productivity than either the U.S. or most LDC's. For one thing, they stressed the development of their human resources by rigorous education, encouraging novel enterprise and stressing quality and human systems management, following the example of Japan. Secondly, they built a good physical and communications infrastructure e.g. credit, banking, telephones, transportation. Thirdly, they chose industries that were well suited to each individual country. Fourth, they provided a stable environment that attracted domestic as well as foreign investment. Does their example suggest the possibility of planned societal change?

It is possible that each society will develop in a unique way, with the world becoming more pluralistic as all societies become industrial and then post-industrial, though at different rates. It is also possible that some societies will remain rooted in their village or river valley cultures indefinitely. Within a generation or two the locus of power and control may, however, shift from the nation-state as a political unit to that of a "region", "community" (e.g. EEC), or "society".

According to K.W. Deutsch, for the time being, only the nation-state can administer to the broadening scope of politics and services. Units that are more localized would fail to get the broad popular support that governments need. And larger units would not provide the services that are sufficiently responsive to the needs of the governed. Technological progress may, however, make the latter more feasible. Moreover, awareness that nation states cannot defend the lives of their citizens may accelerate the shift from nation-states to fewer, larger units or possibly to more, smaller units coordinated and integrated into a coherent world society.

Most probably, part of the world may shift to a few larger units, part to coordinated smaller units, part to uncoordinated smaller units, part to remaining as it is now. Within such a pluralistic world, characterized by increasing economic and political interdependence, conflict and its management may be a paramount concern. The oversimplified dichotomy of information/service orientation on the one hand and ecology/self-service orientation on the other may give way to a new, deeper concept of "information society". It is toward a sketch of this new concept that the remainder of this paper is directed.

Invariants under social transformations

The body of an individual retains characteristic or identifying features by which the individual can be recognized throughout his entire life, despite the continual replacement of most of the cells in his body. Thus, the face, fingerprints, stature, blood type, etc. are invariant from age 20 to 50, or even longer, despite the fact that all the constituents are transformed. The information is genetically encoded as nucleotide sequences in the DNA of each cell or recorded permanently in some other way.

The above are but superficial, physical attributes of an individual. Non-physical attributes, such as predisposition to certain emotional and cognitive disorders as well as some socio-pathologies are increasingly found to have "genetic" origins, though direct DNA encoding for proteins is probably not the mechanism. A greatly broadened concept of "information" is called for. It puts the cognitive component into perspective as a relatively small factor compared with affective and other components.

A society, too, has certain features that remain invariant throughout most of its life. But what are the changes or transformations that it is subject to, especially as it shifts from a pre-industrial to a post-industrial stage? One way to think about these changes is to group them according to the main attributes of the "quality of life" that various scholars seem to agree on. These are: 1) Welfare and survival, measured by mean life expectancy at birth, the distribution of the number of disability-free years over the lifespan, mental well-being, psychic integrity, emotional, or security, etc. This dimension corresponds roughly to what Maslow has called survival and others have called safety and health; 2) Wealth and economic security measured by average per capita purchasing power in that society, total assets or worth per person. It corresponds roughly to what Maslow calls security, what Foa calls money and Chapin economy; it includes tangible liquid and other assets as well as intangible assets; 3) Affection, belongingness, or love, measured approximately by the non-divorce rate, non-crime rate, non-suicide rate, self-esteem, stable reciprocal love relations, acquaintances, etc.; 4) Enlightenment or informedness in the broadest sense, measured by educational level and attainments; it includes ability to choose, know-when know-why, understanding as the recognition of limits to knowledge; 5) Skill, measured by years of experience in jobs requiring special managerial, professional, or technical know-how, as well as basic literacy and numeracy; 6) Rectitude, morality, appreciation, or celebration measured by adherence to religious or ethical rituals, such as church membership, attendance, contributions, but more fundamentally by integration of all personal assets, trustworthiness, honesty, and morality; 7) Power, or the extent to which some persons or groups can enforce their will on others or control who gets what, possibly measured by the number of policies in force that govern or regulate the behavior of many more people than the number of people whose interests these policies explicitly represent; it also includes the ability to mobilize resources according to a strategic plan and deserved authority; 8) Deference or prestige, measured by respect, honor, dignity, reputation, and titles bestowed upon a person (fear may be a component of this as well).

Factor (1) is demographic, while (2) is clearly economic, (3) social, (4) informational, (5) technological, (6) religious, (7) political, and (8) cultural. All these societal aspects change slowly, though there may be momentary surges. Thus, mean life expectancy at birth is increasing slowly in post-industrial societies. It increases more rapidly in LDC's that

TABLE I

Possible Shifts

Dimension	*From present state where:*	*To possible future state where:*
1. Welfare:	The physically strongest live longest and healthiest lives	All who wish it, seek know-how and act on it, can do so too.
2. Wealth:	Those with special privilege, luck, connections, special traits do best	The most knowledgeable do best
3. Affection:	Whose marriage succeeds, whose personality develops into a stable, happy condition seems due to chance	There is more knowledge-based selection of mariage partners, child-rearing, inter-personal & personal adjustment
4. Enlightenment:	Education is oriented to the past and present	Education is oriented to the future
5. Skills:	Work is drudgery, to be done for a living	Skills are used for a pursuit, to achieve something with a passion
6. Rectitude:	Righteousness is ritualistic expedient, uncritical, traditional	Ethical values transcend blind obedience, based on genuine feeling and reason
7. Power:	Those who control physical (e.g. military) resources are on top	Those in rightful authority, by virtue of superior knowledge, are on top
8. Deference:	Honor to the most visible, popular, useful	Honor to the wisest, most deserving by virtue of superior understanding and knowledge that also punctuates practice

introduce modern public health measures to reduce infant mortality and decrease premature mortality from the most prevalent diseases. As a consequence, there is an increasing fraction of the population that is retired or of advanced age in those nations who have passed through the demographic shift and an increased fraction of young in the LDC nations.

Some among these elderly are highly informed, experienced and wise. Their potential is vastly underutilized.

Table I suggests some of the shifts or changes that seem possible along each of the eight dimensions. Admittedly, these shifts have an optimistic, wishful component. An equally valid scenario would replace the assertions in the last column by those of the opposite meaning, suggesting that the future is "worse" than the present.

The driving forces underlying these changes, especially in LDC's, are continued population growth and the possibility of involvement in technologies such as microelectronics, robots, computerized information systems, genetic engineering products, new kinds of materials, etc. Whether the shifts are in the direction a society considers desirable or undesirable depends on its values and its human management. Regardless of what changes do occur, some aspects of these 8 dimensions remain unchanged for each surviving society.

Table II is comprised of mere guesses by the author at what these unchanging cultural patterns are in one American society. The point of this section is to stress that for a society to be an information society it is necessary (though not sufficient) that information about those invariants is recorded in its structural memory, circulated freely among its members and used in selecting and pursuing societal goals and making choices.

All of these invariants are in the form of consensually held beliefs that reflect basic values. A society mobilizes to fight if these values are threatened. Often, however, it becomes aware of these invariants only

TABLE II

Dimensions	*Possible Invariants Under Change*
1. Welfare	Belief in Sanctity of Individual Life. No person is dispensible
2. Wealth	Belief in Equality of Opportunity
3. Affection	Belief in golden rule (do unto others as you would have them do unto you), reaching out, social justice
4. Enlightenment	Humanistic, liberal education of the whole person in enduring values and rational thought
5. Skills	Success depends on effort and ability, not luck or position, and effort should be worthwhile, justly rewarded
6. Rectitude	Judaeo-Christian (Protestant) ethic, e.g.
7. Power	Individual political freedoms; supremacy of electorate
8. Rectitude	Honor whom honor is due based on values of justice, fairness, truth, goodness, beauty

when they are threatened by a major external challenge as in the case of fascism or communism for the U.S. society. In the absence of outside shocks, the invariants are not always in the public consciousness nor even in that of its leaders. This is apparent when choices must be made.

The idea of societal invariants is, of course, at the core of the "society" concept. The traditional definition of "human society" refers to a group of persons sharing a "self-sufficient" system of action that lasts longer than its members; internal organization, territoriality, cultural distinctiveness, and sexual recruitment are often used to characterize a "society". These, as well as provisions for adequate relations with the environment, role differentiation and allocation of members to roles, a communication infrastructure, a shared view of the world, societal goals, and consensual norms exhibited in behavior for purposes of socialization and social control (e.g. of deviants) are mechanisms to ensure self-maintenance and non-termination due to biological extinction, dispersion, apathy, assimilation, or disintegration. It is not clear whether or not a group, to be "society", must also be able to employ force against outsiders and to limit use of force within the group. If so, then many nation-states are societies, though many tribes or clans would be, as well. In a non-information society in this sense, a system of action is shared without that being conscious and symbolically encoded and circulated through the social fabric. Communication networks exist. Affect as well as some information flows through the social bonds of such a society, but affective and social rather than cognitive aspects of information predominate. In an information society that is reversed. The cognitive aspects predominate, though they must be integrated and consistent with the affective and social aspects.

The introduction of technologies, appropriate or not, increases the number of options available to the members of a society. Thus, the introduction of a vaccine against a major illness and means for mass innoculations gives everyone a choice they did not have before. By freeing many of the disease and possibly premature death, it also gives them many more choices about how to spend their lives. Yet, adherents to certain religious cults might refuse vaccination in the belief that it interferes with God's will; if such a belief is an invariant in that society, it governs the choice among options.

It does not follow that each society in a culturally pluralistic world, having identified, encoded, and recorded its societal invariants should or would indefinitely adhere to these in as rigid a way as they now adhere to time-honored traditions, constitutional principles, ideological dogmas, or religious doctrines. Increased interdependence and interaction of all kinds among the world's new societal units may stimulate cultural diffusion and change, call into question traditional beliefs and values. The constant interplay between what is to remain invariant and what should adapt is characteristic of a dynamic information society. Its management of these inevitable conflicts by "information technologies" (i.e. reason and knowledge) rather than by traditional means is what makes it an information society.

We now present the proposed new concept of information society. An *information society* is one that: (1) enables most of its members to engage in productive pursuits that are knowledge-intensive, knowledge-generating, and knowledge-based; (2) has a communication network that freely circulates information about its basic societal invariants so that this information is consistently, effectively, and efficiently acted upon in the making of choices; (3) manages inevitable conflict between conservative pressures to preserve these invariants and pressures for adaptive change by reason, knowledge-based understanding, and enlightened creative wisdom blended with human values, rather than traditional resort to base emotions and brute force alone.

In what follows, we will try to explicate and justify this concept and analyze its implications. In the above, the term "knowledge" in its broadest sense is intended in place of "information." This essay's broad cultural thrust is related to the intensive economic analysis of the information discipline, particularly the labor implications – condition (1) above – by adding conditions (2) and (3).

Inputs, production, and outputs in an information society

We begin with the first of the three conditions we view as necessary for a society to be an information society. It concerns the economic aspects. Production, distribution, and consumption of goods and services depend increasingly in every society on a literate and enlightened population. Technologies now available to every society are reducing the need for human labor requiring physical strength and great dexterity. This now is increasingly the case for labor requiring routine mental operations, such as book-keeping. But a reduction of the need for certain kinds of human labor in a society does not imply an actual reduction in such a labor force. Organized labor or governments may erect protectionist barriers. More importantly, the basic societal invariants come into play.

It would be facile to define an information society as one in which the means of production are primarily based on information technology and both the inputs and outputs are primarily information. An example would be the production of health-related advice using computer-based expert knowledge systems, the inputs being medical know-how and the outputs medical advice tailored to the needs of a user. In some societies, such an automated medical advisory service might be widely accepted if it is consistent with the beliefs inherent in that society's invariants. It is less likely to be accepted in a society that values personal, face-to-face interactions and the *means* of communication as much as its content. In a highly context-sensitive society, such as Japan, one-dimensional approaches to many-dimensional situations would not gain acceptance. While a society in which such information technologies are appropriate in that sense is properly called an information society, a society in which such a technology is introduced when it is inappropriate or unnatural should not be called an information society.

Productive activities in which extensive and up-to-date know-what, know-whom, know-how much, know-why, or know-when is necessary to carry them out shall be called *knowledge-based*. That includes nearly all professions, such as engineering, law, medicine, architecture, etc. Crafts are not knowledge-intensive. A specific skill is not know-how, except for that one skill; know-how involves a repertoire of skills and the knowledge to choose appropriately.

Productive activities in which knowledge is the limiting input factor will be called *knowledge-intensive*. Ministering to the sick is an example because the type of treatment depends critically on information about the patient's condition. Without enough information that is also good enough, the activity can harm as much as it can help. But there can be knowledge-intensive activities that are not knowledge-based, as is the case for a guard with the single skill of distinguishing friend from foe and stopping the latter. There can also be knowledge-based activities that are not knowledge intensive, as in the case of a chef who must have a great deal of know-how but need not depend on information inputs.

If the output of a productive activity is primarily an aspect of new knowledge, we call it *knowledge-generating*. Not every knowledge-intensive activity is knowledge-generating. A physician may acquire knowledge to treat a patient; in doing so he generates knowledge about the patient's probable condition, but that is only a means to the end of treating the patient. Diagnostic knowledge for routine conditions is a secondary rather than a primary output. On the other hand, knowledge-generating activities that are not knowledge-intensive may also be possible. The sculptor's or composer's output is primarily a form of knowledge, but no information input is a limiting factor.

Knowledge-generating activities need not be knowledge-based, as illustrated by the work of a copier, and knowledge-based activities need not result in knowledge-generation, as illustrated by the work of a healer. Some occupations, such as legal services, are all three: (1) knowledge-intensive (knowledge of law, and the client's problem, are limiting factors), (2) knowledge-based (knowing how to draft contracts, what to put in them, who the adversary attorneys or judges are likely to be, how much to include, leave out, ask for in a suit, charge, when to sue or not to sue and being able to justify all steps); and (3) knowledge-generating (producing legal advice, a new contract, a judicial decision, etc.) all at once. An occupation such as law, in which most activities are all three, will be called a *knowledge occupation*. A society in which at least half of its members are productively engaged in knowledge occupations is the first condition for a society to be called an information society.

This definition is considerably more restrictive than that in current use. Some scholars require workers in knowledge occupations to perform activities that are knowledge-generating only, regardless of whether they are also knowledge-based and knowledge-intensive. They do not require that it maintain and use information about societal invariants or use information technologies for conflict management. On the other hand, they include as knowledge activities crafts, trades, unskilled work, and professional work – done primarily to earn money – in such fields as insurance, banking, education, etc. We propose to broaden the concept of "occupation" to include what F. Bergmann calls a pursuit: activities done with a passion and not necessarily or primarily to earn money – or what P. Sonntag calls a productive hobby. That includes the "unpaid overtime" activities of many scientists, artists, teachers, scholars, politicians, clergymen, and a few in the healing arts.

The distinction between work and pursuit is important because the accelerated pervasiveness of current technologies, which will make their main impacts on society only in the next decade or two, should vastly increase the productivity of human labor at all levels. Unless demand for new goods and services is generated as fast as this ability to meet it with less labor, the total demand for paid labor should decrease. Available funds for human labor must either be distributed among more people (shorter and shorter work-hours) or we should expect widespread structural unemployment. Productive persons will, however, not be content without utilizing the "unpaid" time thus made available to them without attempting to develop themselves and their productive potential more fully. This will lead them to pursuits, probably in knowledge occupations as we defined them, and toward a new kind of market in the new knowledge that they generate, as in Masuda's vision of information society. To maintain information about societal invariants in the consciousness of all segments of the society at all times would not be possible for a large, geographically dispersed population that is aware of the options of modern life without a modern information technology. The media of print, radio, television, paper or blackboards to write on, screens on which to project overhead transparencies, etc. have for some time been indispensable tools for education, advertising, and entertainment, to mention but three routes by which information about societal invariants is disseminated to masses of the population, and through which consensual values and belief systems are shaped and maintained. Copying machines and, currently, computer-communication networks have the potential for making those routes vastly more effective and of opening new and more powerful routes.

It is noteworthy that in societies where power is centralized and which subordinate the goals and needs of individuals to those of the society as a unit, as conceived by those in power, the use of copiers and computer terminals tied to networks is allowed only to those in the power elite. In the U.S.S.R., for example, the lens of a copying machine in a research institute is kept in a separate locked room from that of the machine so that unauthorized uses are very difficult.

Exchange and control

In Masuda's vision, members of an information society will exchange the fruits of their knowledge-generating activities directly.

They will, of course, still have to pay their bills for conventional goods and services by conventional price/money-based market mechanisms, even if information technologies in the future replace cash, checks, and credit cards by electronic funds transfer methods. The notion of a market, which is inherently an information system, may, however, undergo important revisions. Current information technologies are already creating new definitions of traditional markets and distribution channels. Persons in need of certain goods and services, who would admit, if asked, to seeking to obtain highest quality at lowest available prices, would be far better informed about *all* the available options in quality and price. They would also need a secondary information system to help them assimilate and choose from a possibly overwhelming and bewildering amount of such information. Vendors, too, would be far better informed than they are now about consumer needs and their likely responses to the market positions of their goods and services.

Marketing and exchange of knowledge products and services, however, may differ. Such outputs differ basically from materials or energy-based outputs in that they are not consumed. On the contrary, their use may make them more visible, more wanted, and more available. The concepts of ownership and control of knowledge outputs may change. Discoveries and inventions, though generally made independently and within the same decade or two by several persons or groups, are increasingly recognized as the result of social or communal activities, even if they are still attributed to individuals. Information can be assigned economic value, as in the case of that produced by a market survey for a company that caused management to reverse a decision to produce some product with a resulting profit. That same information, however, has the same value to that company's competitors. Unlike other goods and services, the quantity of such information demanded and supplied may not vary with price, though if there are several market surveyors able to produce the same results, that will keep the price down.

The intellectual market for the free exchange of ideas, insights, questions-and-answers, and accounts of experiences may assume, in information societies, an importance equivalent to the market for goods and services. This is already the case in some scientific and artistic communities, though only by means of face-to-face encounters, professional journals, and the telephone. The potentials of computer conferencing and two-way interactions cable television are just now beginning to be developed.

More educated people with more time to reflect, to realize the choices available to them, and the difficulties of choosing wisely may pursue both the search for their society's invariants at the same time that they seek self-actualization and harmony with their local environment. In an information society they are able to use computers to manage inner conflict and to use their on-line access to networks for management of interpersonal conflict.

At the same time some North American, Japanese, and European communities become information societies, close to two billion persons under age twenty will have to be assimilated into pre-industrial societies. The basic needs of billions in these societies could and should be the economic demand for conventional goods and services that industrial and post-industrial societies can supply and must supply if they are to operate at full capacities and employment in conventional economic activities.

Needs do not become economic demand without purchasing power. That can be achieved only by exporting items of value that they control or develop the ability to produce. That takes know-how and it may be in the interest of the information societies to help the other societies develop the appropriate know-how.

The growing interdependence of nation-states and the possibility that in a generation or two nation-states may no longer be the only or the dominant political unit may support the belief of some in the existence of an "emerging globalism." That is assumed to have three components: (1) information technologies; (2) ecological attitudes; (3) "mass individuals". Masses of people are seen to form information societies dependent on special relationships with their sustaining ecosystems. Database management systems, teleconferencing, "New Alchemy Arks for Survival", global environmental monitoring, computerized life-support crisis management, etc. are suggested means of effecting the needed integration.

We have tried to suggest, throughout this essay, that a consensual belief about a desirable (or undesirable) society sometimes functions like a self-fulfilling prophecy. Consensus is mediated by an information network akin to the nervous system of an individual, except that it must be even more complex. It is therefore reasonable to make the existence of such an information network, evidenced by the circulation of information about societal invariants, a second necessary condition for an *information* society. We would call a post-industrial society, such as the U.S., an information society when information and knowledge began to appear as explicit entries in national or societal income and expenditure accounts, and when information is recognized explicitly as an output.

Finally, "information" in an information society is not an end in itself, but a means, much the same as material goods and services. The end may be a quest, perhaps one that is never attainable, but one that surviving societies approach, for the "mystical experience of flow and wholeness," or a kind of modern spiritualism that integrates and synthesizes our cognitive/scientific, affective, conative, and moral knowledge into wisdom. This kind of synthesis, at higher levels of consciousness, was meant to be captured by the third requirement for an information society.

The three conditions in the proposed new concept of "information society" make distinctions that are more appropriate than the existing ones. The growth issue is replaced by concern with the shift from material goods and physical energy to knowledge/information. In the latter, there appear to be no limits to growth. The issue of affluence vs. frugality is replaced by the consideration that information technologies may bring affluence to most of the world, and even eliminate the need for routine, physical labor by persons, while meshing in optimum synchrony with the natural environment. Control of complexity is an issue, but one that the introduction of *human* systems management is likely to cope with. As Catton puts it: "The paramount need of post-exuberant humanity is to remain human in the face of dehumanizing pressures."

An information society, then, is a stage in the evolution of "community brains", toward a "world brain". This is probably most likely to be the essence of the "great transition" that futurists seem to agree on. When enough people begin to believe it, as is likely to happen if it is a stage in natural cultural evolution, then this belief may contribute to its self-fulfillment – perhaps one of nature's evolutionary mechanisms. It will still take some decades before this idea is sufficiently widespread, and until the first information society (in the sense defined here) appears. If this paper has spread the idea to a few, it will have achieved its purpose.

Conclusion

It is possible to imagine in the future a pluralistic, decentralized world in which some post-industrial nation-states evolve into information societies characterized by three features. First, there is a preponderance of activities that are knowledge-generating, knowledge-based, and knowledge intensive. Second, they have a communication net in which they circulate, store and use information about their societal invariants, their unchanging values, and beliefs. Third, they use information technologies to manage conflicts between what is to be invariant and what is to be planned adaptive change. They direct as many of these activities as necessary to develop know-how for the creation of purchasing power in less-developed countries which they share freely with them.

The brain organization of the modern world

H. G. Wells

[...] I know of no general history of human education and discussion in existence. We have nowadays in what is called the New History – books which trace for us in rough outline the growth in size and complexity of organized human communities. But so far no one has attempted to trace the stages through which *teaching* has developed, how schools began, how discussions grew, how knowledge was acquired and spread, how the human intelligence kept pace with its broadening responsibilities. We know that in the small tribal community and even in the city states of, for example, Greece, there was hardly any need for reading or writing. The youngsters were instructed and initiated by their elders. They could walk all over the small territory of their community and *see* and *hear*, how it was fed, guarded, governed. The bright young men gathered for oral instruction in the Porch or the Academy. With the growth of communities into states and kingdoms we know that the medicine man was replaced by an organized priesthood, we know that scribes appeared, written records. There must have been schools for the priests and scribes, but we know very little about it. We know something of the effect of the early writings, the Bible particularly, in consolidating and preserving the Jewish tradition – giving it such a start off that for a long time it dominated the subsequent development of the Gentile world, and we know that the survival and spread of Christianity is largely due to its resort to written records to supplement that oral teaching of disciples with which it began. But the growing thirst for medical, theological and general knowledge that appeared in the Middle Ages and which led to those remarkable gatherings of hungry minds, the universities, has still to be explained and described. That appearance and that swarming of scholars would make an extraordinary story. After the lecture room, the book; after that the newspaper, universal education, the cinema, the radio. No one has yet appeared to make an orderly story of the developments of information and instruction that have occurred in the past hundred years. Age by age the World's Knowledge Apparatus has grown up. Unpremeditated. Without a plan. But enlarging the possible areas of political co-operation at every stage in its growth.

It is a very interesting thing indeed to ask oneself certain questions. How did I come to know what I know about the world and myself? What ought I to know? What would I like to know that I don't know? If I want to know about this or that, where can I get the clearest, best and latest information? And where did these other people about me get their ideas about things? Which are sometimes so different from mine. Why do we differ so widely? Surely about a great number of things upon which we differ there is in existence exact knowledge? So that we ought not to differ in these things. This is true not merely about small matters in dispute but about vitally important things concerning our business, our money, our political outlook, our health, the general conduct of our lives. We are guessing when we might know. The facts are there, but we don't know them completely. We are inadequately informed. We blunder about in our ignorance and this great ruthless world in which we live, beats upon us and punishes our ignorance like a sin. Not only in our mass-ruled democracies but in the countries where dogmas and dictators rule, tremendous decisions are constantly being made affecting human happiness, root and branch, in complete disregard of realities that are *known*.

You see we are beginning to realize not only that the formal political structures of the world and many of the methods of our economic life are out of date and out of scale, but also another thing that hampers us hopelessly in every endeavour we make to adjust life to its new conditions – our World Knowledge Apparatus is not up to our necessities. We are neither collecting, arranging nor digesting what knowledge we have at all adequately, and our schools, our instruments of distribution are old-fashioned and ineffective.

We are not being told enough, we are not being told properly, and that is one main reason why we are all at sixes and sevens in our collective life.

The other day my university, the University of London, celebrated its centenary. For some minor reason I was asked to assist at these celebrations. And to do so I had to assume some very remarkable garments – most remarkable if you consider that London University was founded in the year 1836 when gentlemen wore tight trousers with straps, elegantly waisted coats and bell shaped top hats. Did I dress up like that? No. I found myself retreating from the age of the aeroplane to the age of the horse and mule outfit of the Canterbury Pilgrims. I found myself wearing a hood and gown and carrying a Beret rather like those worn by prosperous citizens of the days of Edward IV, when the University of London was as little anticipated as the continent of America. My modern head peeped out at the top of this get-up and my modern trousers at the bottom. Properly I ought to have been wearing a square beard or have been clean-shaven, but I was forgiven that much. And from all parts of the world representatives of innumerable universities had come with beautifully illuminated addresses to congratulate our Chancellor and ourselves on our hundred years of sham mediaevalism. They came from the ends of the earth, they came up the aisle in an endless process; one ancient name followed another, now it was Tokyo, now Athens, now Uppsala, now Cape Town, now the Sorbonne, now Glasgow, now Johns Hopkins, on they came and on and bowed and handed their addresses and passed aside. It was a marvellous, a dazzling array of beautifully coloured robes. It was also a marvellous collection of men and women. I watched the grave and dignified faces of some of the finest minds in the world. Together they presented, they embodied or they were there to represent, the whole body of human knowledge. There it was in effect parading before me. And nine out of ten of them were dressed up in some colourful imitation of a costume worn centuries before their foundations came into existence. It was picturesque, it was imposing – but it was just a little odd of them.

My thoughts drifted away to certain political gatherings I had seen and heard; faces of an altogether inferior type, leather-lunged adventurers bawling and gesticulating, raucous little men screaming plausible nonsense to ignorant crowds, supporters herded like sheep and saluting like trained monkeys, and the incongruity of the contrast came to me – you know how things come to you suddenly at times – so that I almost laughed aloud. Because, when it comes to the direction of human affairs, all these universities, all these nice refined people in their lovely gowns, all this visible body of human knowledge and wisdom, has far less influence upon the conduct of human affairs, than, let us say, an intractable newspaper proprietor, an unscrupulous group of financiers or the leader of a recalcitrant minority.

Some weeks previously I had taken part in a little private conference of scientific men in London. They were very distinguished men indeed, and they were distressed beyond measure at the way in which one scientific invention after another was turned to the injury of human life. What was to be done? What could be done? Our discussion was inconclusive, but it had quickened my sense of the reality of the situation. I put these three separate impressions together before you: first, these anxious scientific specialists, then the unchallenged power and mischief of these bawling war-making politicians and their crowds at the present time, and finally, capping the whole, these hundreds of all-too-decorated learned gentlemen, fine and delicate, bowing, presenting addresses (for the most part in Latin) and conferring further gowns and diplomas on one another. This last lot, I said, this third lot is after all – in spite of its elegant weakness, the organized brain of mankind so far as there is an organized brain of mankind – and it is not doing its proper work. Why? Why are our universities floating above the general disorder of mankind like a beautiful sunset over a battlefield? Is it not high time that something was done about it?

Certain ideas had been stirring in my mind for some time already, but this scene of archaic ceremony just lit up the situation for me. I realized that these mediaeval robes were in the highest degree symptomatic. They clothed an organization essentially mediaeval, inadequate and out of date. We are living in 1937 and our universities, I suggest, are not half-way out of the fifteenth century. We have made hardly any changes in our conception of university organization, education, graduation, for a century – for several centuries. The three or four years' course of lectures, the bachelor who knows some, the master who knows *most,* the doctor who knows *all,* are ideas that have come down unimpaired from the Middle Ages. Nowadays no one should end his learning while he lives and these university degrees are preposterous. It is true that we have multiplied universities greatly in the past hundred years, but we seem to have multiplied them altogether too much upon the old pattern. A new battleship, a new aeroplane, a new radio receiver is always an improvement upon its predecessor. But a new university is just another imitation of all the old universities that have ever been. Educationally we are still for all practical purposes in the coach and horse and galley stage. The new university is just one more mental *gilt-coach* in which minds take a short ride and get out again. We have done nothing to co-ordinate the work of our universities in the world – or at least we have done very little. What are called the learned societies with correspondents all over the world have been the chief addition to the human knowledge organization since the Renaissance and most of these societies took their shape and scale in the eighteenth and nineteenth centuries. All the new means of communicating ideas and demonstrating realities that modern invention has given us, have been seized upon by other hands and used for other purposes; these universities which should guide the thought of the world, making no protest. The showmen got the cinema and the governments or the adventurers got the radio. The university teacher and the schoolmaster went on teaching in the class-room and checking his results by a written examination. It is as if one attempted to satisfy the traffic needs of greater New York or London or Western Europe by a monstrous increase in horses and carts and nothing else. The universities go out to meet the tremendous challenges of our social and political life, like men who go out in armour with bows and arrows to meet a bombing aeroplane. They are pushed aside by men like Hitler, Mussolini creates academies in their despite, Stalin sends party commissars to regulate their researches. It is beyond dispute that there has been a great increase in the research work of universities; that pedantry and mere scholarship in spite of an obstinate defence have declined relatively to keen inquiry, but the specialist is by his nature a preoccupied man. He can increase knowledge, but without a modern organization backing him he cannot put it over. He can increase knowledge which ultimately is power, but he cannot at the same time control and spread this power that he creates. It has to be made generally available if it is not to be monopolized in the wrong hands.

There, I take it, is the gist of the problem of World Knowledge that has to be solved. A great new world is struggling into existence. But its struggle remains catastrophic until it can produce an adequate knowledge organization. It is a giant birth and it is mentally defective and blind. An immense and ever-increasing wealth of knowledge is scattered about the world today, a wealth of knowledge and suggestion that – systematically ordered and generally disseminated – would probably give this giant vision and direction and suffice to solve all the mighty difficulties of our age, but that knowledge is still dispersed, unorganized, impotent in the face of adventurous violence and mass excitement. In some way we want to modernize our World Knowledge Apparatus so that it may really bring what is thought and known within reach of all active and intelligent men. So that we shall know – with some certainty. So that we shall not be all at sixes and sevens about matters that have already been thoroughly explored and worked out.

How is that likely to be done?

Not of course in a hurry...

It would be very easy to do a number of stupid things about it – futile or even disastrous things. I can imagine quite a number of obvious preposterous mischievous experiments, a terrible sort of world university consolidation, an improvised knowledge dictatorship. Heaven save us from that! We want nothing that will in any sense override the autonomy of institutions or the independence of individual intellectual workers. We want nothing that will invade the precious time and attempt to control the resources of the gifted individual specialist. He is too much distracted by elementary teaching and college administration already. We do not want to magnify and stereotype universities. Most of them with their gowns and degrees, their slavish imitation of the past, are too stereotyped already.

But here it is that the idea I want to put before you comes in, this idea of a greater encyclopaedism – with a permanent organism and a definite form and aim. I put forward the development of this new encyclopaedism as a possible method, the only possible method I can imagine, of bringing the universities and research institutions of the world into effective co-operation and creating an intellectual authority sufficient to control and direct our collective life. I imagine it as a permanent institution – untrammelled by precedent, a new institution – something *added* to the world network of universities, linking and co-ordinating them with one another and with the general intelligence of the world. Manifestly as my title for it shows, it arises out of the experience of the French Encyclopaedia, but the form it is taking in the minds of those who have become interested in the idea, is of something vastly more elaborate, more institutional and far-reaching than Diderot's row of volumes. The immense effect of Diderot's effort in establishing the frame of the progressive world of the nineteenth century, is certainly the inspiration of this new idea. The great rôle played in stabilizing and equipping the general intelligence of the nineteenth century world by the French, the British and the German and other encyclopaedias that followed it, is what gives confidence and substance to this new conception. But what we want today to hold the modern mind together in common sanity is something far greater and infinitely more substantial than those earlier encyclopaedias. They served their purpose at the time, hut they are not equal to our current needs. A World Encyclopaedia no longer presents itself to a modern imagination as a row of volumes printed and published once for all, but as a sort of mental clearing house for the mind, a depot where knowledge and ideas are received, sorted, summarized, digested, clarified and compared. It would be in continual correspondence with every university, every research institution, every competent discussion, every survey, every statistical bureau in the world. It would develop a directorate and a staff of men of its own type, specialized editors and summarists. They would be very important and distinguished men in the new world. This Encyclopaedic organization need not be concentrated now in one place; it might have the form of a network. It would centralize mentally but perhaps not physically. Quite possibly it might to a large extent be duplicated. It is its files and its conference rooms which would be the core of its being, the essential Encyclopaedia. It would constitute the material beginning of a real World Brain.

Then from this centre of reception and assembly, would proceed what we may call the Standard Encyclopaedia, the primary distributing element, the row of volumes. This would become the common backbone as it were of general human knowledge. It might take the form of twenty or thirty or forty volumes and it would go to libraries, colleges, schools, institutions, newspaper offices, ministries and so on all over the world. It would be undergoing continual revision. Its various volumes would be in process of replacement, more or less frequently according to the permanence or impermanence of their contents. And from this Standard Encyclopaedia would be drawn a series of text-books and shorter reference encyclopaedias and encyclopaedic dictionaries for individual and casual use.

That crudely is the gist of what I am submitting to you. A double-faced organization, a perpetual digest and conference on the one hand and a system of publication and distribution on the other. It would be a clearing house for universities and research institutions; it would play the rôle of a cerebral cortex to these essential ganglia. On the one hand this organization should be in direct touch with all the original thought and research in the world; on the other it should extend its informing tentacles to every intelligent individual in the community – the new world community.

In that little world of the eighteenth century, what we may call the mind of the community scarcely extended below the gentlefolk, the clergy and the professions. There was no primary education for the common man at all. He did not even read. He was a mere toiler. It hardly mattered how little he knew – and the less he thought the better for social order. But machinery abolishes mere toil altogether. The new world has to consist of a world community – say of 2,000 million educated individuals – and the influence of the central encyclopaedic organization, informing, suggesting, directing, unifying, has to extend to every rank of society and to every corner of the world. The new encyclopaedism is merely the central problem of world education.

Perhaps I should explain that when I speak in this connexion of universities, what I have in mind is primarily assemblies of learned men or men rehearsing their ripe scholarship or conducting original research with such advanced students and student helpers as have been attracted by them and are sharing their fresh and inspiring thoughts and methods. This is a return to the original university idea. The original universities were gatherings of eager people who wanted to know – and who clustered round the teachers who did seem to know. They gathered about these teachers because that was the only way in which they could get their learning. I am talking of that sort of university. That is the primary form of a university. I am not talking here of the collegiate side of a contemporary university, the superficial finishing school exercises of sportive young people mostly of the wealthier classes who don't want to know – young people who mean very little and who have been sent to the university to make useful friendships and get pass-degrees that mean hardly anything at all. These mere finishing-school students are a modern addition, a transitory encumbrance of the halls of learning. I suppose that before very long much of this undergraduate life will merge with the general upward extension of educational facilities to all classes of the community. I assume that the tentacles of this Encyclopaedia we are anticipating, with its comprehensive and orderly supply of knowledge, would intervene beneficially between the specialized research and learning which is the living reality in the university and this really quite modern finishing-school side. The time is rapidly returning when men of outstanding mental quality will consent to teach only such students as show themselves capable of and willing to follow up their distinctive work. The mere graduating crowd with their games and their yells and so forth, will go back to the mere teaching institutions where they properly belong.

But I will not spend the few minutes remaining to me upon which is after all a side issue in this discussion. University organization is not now my subject. I am talking of an essentially *new* organization – an *addition* to the intellectual apparatus of the world. The more important thing now is to emphasize this need – a need the world is likely to realize more and more acutely in the coming years – for such a concentration, which will assemble, co-ordinate and distribute accumulated knowledge. It will link, supplement and no doubt modify profoundly, the universities, schools and other educational organizations we possess already, but it will not in itself be a part of them.

Let me make it perfectly clear that for the present it is desirable to leave this project of a World Encyclopaedic organization vague – in all but its essential form and function. It might prove disastrous to have it crystallize out prematurely. Such premature crystallization of a thing needed by the world can produce, we now realize, a rigid obstructive *reality*, just *like enough* to our actual requirements to cripple every effort to replace it later by a more efficient organization. Explicit constitutions for social and political institutions, are always dangerous things if these institutions are to live for any length of time. If a thing is really to live it should grow rather than be made. It should never be something cut and dried. It should be the survivor of a series of trials and fresh beginnings – and it should always be amenable to further amendment.

So that while I believe that ultimately the knowledge systems of the world must be concentrated in this world brain, this permanent central Encyclopaedic organization with a local habitat and a world-wide range – just as I believe that ultimately the advance of aviation *must* lead, however painfully and tortuously by way of World Air Control, to the political, economic and financial federation of the world – yet nevertheless I suggest that to begin with, the evocation of this World Encyclopaedia may begin at divergent points and will be all the better for beginning at divergent points.

Of the demand for it, and of the readiness for it in our world today, I have no sort of doubt. Ask the book-selling trade. Any books that give or even seem to give, any sort of conspectus of philosophy, of science, of general knowledge, have a sure abundant sale. We have the fullest encouragement for bolder and more strenuous efforts in the same direction. People *want* this assembling of knowledge and ideas. Our modern community is mind-starved and mind-hungry. It is justifiably uneasy and suspicious of the quality of what it gets. The hungry sheep look up and are not fed – at least they are not fed properly. They want to know. One of the next steps to take, it seems to me, is to concentrate this diffused demand, to set about the definite organization of a sustained movement, of perhaps a special association or so, to bring a World Encyclopaedia into being. And while on the one hand we have this world-wide receptivity to work upon, on the other hand we have among the men of science in particular a very full realization of the need for a more effective correlation of their work. It is not only that they cannot communicate their results to the world; they find great difficulty in communicating their results to one another. Among other collateral growths of the League of Nations is a certain Committee of Intellectual Co-operation which has now an official seat in Paris. Its existence shows that even as early as 1919, someone had realized the need for some such synthesis of mental activities as we are now discussing. But in timid, politic and scholarly hands the Committee of Intellectual Co-operation has so far achieved little more than a building, a secretary and a few salaries. The bare idea of a World Encyclopaedia in its present delicate state would give it heart failure. Still there it is, a sort of seed that has still to germinate, waiting for some vitalizing influence to stir it to action and growth. And going on at present, among scientific workers, library workers, bibliographers and so forth, there is a very considerable activity for an assembling and indexing of knowledge. An important World Congress of Documentation took place this August in Paris. I was there as an English delegate and I met representatives of forty countries – and my eyes were opened to the very considerable amount of such harvesting and storage that has already been done. From assembling to digesting is only a step – a considerable and difficult step but, none the less, an obvious step.

In addition to these indexing activities there has recently been a great deal of experimentation with the microfilm. It seems possible that in the near future, we shall have microscopic libraries of record, in which a photograph of every important book and document in the world will be stowed away and made easily available for the inspection of the student. The British Museum library is making microfilms of the 4,000 books it possesses that were published before 1550 and parallel work is being done here in America. Cheap standardized projectors offer no difficulties. The bearing of this upon the material form of a World Encyclopaedia is obvious. The general public has still to realize how much has been done in this field and how many competent and disinterested men and women are giving themselves to this task. The time is close at hand when any student, in any part of the world, will be able to sit with his projector in his own study at his or her convenience to examine *any* book, *any* document, in an exact replica.

Concurrently with this movement towards documentation, we may very possibly have a phase when publishers will be experimenting in the production of larger and better Encyclopaedias, all consciously or unconsciously attempting to realize the final world form. And satisfy a profitable demand. The book salesman from the days of Diderot onward has shown an extraordinary knack for lowering the quality of this sort of enterprise, but I did not see why groups of publishers throughout the world should not presently help very considerably in the beginning of a permanent Encyclopaedic foundation. But such questions of ways and means of distribution belong to a later stage of this great intellectual development which lies ahead of us. I merely glance at them here.

There are certain responses that I have observed crop up almost automatically in people's minds when they are confronted with this project of a world-wide organization of all that is thought and known. They will say that an Encyclopaedia must always be tendentious and within certain limits – but they are very wide limits – that must be true. A World Encyclopaedia will have by its very nature to be what is called *liberal*. An Encyclopaedia appealing to all mankind can admit no narrowing dogmas without at the same time admitting corrective criticism. It will have to be guarded editorially and with the utmost jealousy against the incessant invasion of narrowing propaganda. It will have a general flavour of what many people will call scepticism. Myth, however venerated, it must treat as myth and not as a symbolical rendering of some higher truth or any such evasion. Visions and projects and theories it must distinguish from bed-rock fact. It will necessarily press strongly against national delusions of grandeur, and against all sectarian assumptions. It will necessarily be *for* and not indifferent to that world community of which it must become at last an essential part. If that is what you call bias, bias the world Encyclopaedia will certainly have. It will have, and it cannot help but have, a bias for organization, comparison, construction and creation. It is an essentially creative project. It has to be the dominant factor in directing the growth of a new world.

Well, there you have my anticipation of the primary institution which has to appear if that world-wide community towards which mankind, willy-nilly, is being impelled, is ever to be effectively attained. The only alternative I can see is social dissolution and either the evolution of a new, more powerful type of man, or the extinction of our species. I sketch roughly – it seems to be my particular rôle in life to do these broad sketches and outlines and then stand aside – but I do my best to make the picture plain and understandable. And for me at any rate this is no Utopian dream. It is a forecast, however inaccurate and insufficient, of an absolutely essential part of that world community to which I believe we are driving now. I do not believe there is any emergence for mankind from this age of disorder, distress and fear in which we are living, except by way of such a deliberate vast reorganization of our intellectual life and our educational methods as I have outlined here. I have been talking of real intellectual forces and foreshadowing the emergence of a vital reality. I have been talking of something which may even be recognizably in active operation within a life time – or a lifetime or so, from now – this consciously and deliberately organized brain for all mankind.

In a few score years there will be thousands of workers at this business of ordering and digesting knowledge where now you have one. There will be a teacher for every dozen children and schools as unlike the schools of today as a liner is unlike the *Mayflower*. There will not be an illiterate left in the world. There will hardly be an uninformed or misinformed person. And the brain of the whole mental network will be the Permanent World Encyclopaedia.

Well, I have designedly put much controversial matter before you, and I have not hesitated to put it in a provocative manner. You will, I know, understand that every new thing is apt to seem crude at first. Forgive my crudities. But my time has been short for what I had to say, and I have said it in the way that seemed most challenging and most likely to produce further discussion.

As we may think

Vannevar Bush

[...]
There is a growing mountain of research. But there is increased evidence that we are being bogged down today as specialization extends. The investigator is staggered by the findings and conclusions of thousands of other workers – conclusions which he cannot find time to grasp, much less to remember, as they appear. Yet specialization becomes increasingly necessary for progress, and the effort to bridge between disciplines is correspondingly superficial.

Professionally our methods of transmitting and reviewing the results of research are generations old and by now are totally inadequate for their purpose. If the aggregate time spent in writing scholarly works and in reading them could be evaluated, the ratio between these amounts of time might well be startling. Those who conscientiously attempt to keep abreast of current thought, even in restricted fields, by close and continuous reading might well shy away from an examination calculated to show how much of the previous month's efforts could be produced on call. Mendel's concept of the laws of genetics was lost to the world for a generation because his publication did not reach the few who were capable of grasping and extending it; and this sort of catastrophe is undoubtedly being repeated all about us, as truly significant attainments become lost in the mass of the inconsequential.

The difficulty seems to be, not so much that we publish unduly in view of the extent and variety of present-day interests, but rather that publication has been extended far beyond our present ability to make real use of the record. The summation of human experience is being expanded at a prodigious rate, and the means we use for threading through the consequent maze to the momentarily important item is the same as was used in the days of square-rigged ships.

But there are signs of a change as new and powerful instrumentalities come into use. Photocells capable of seeing things in a physical sense, advanced photography which can record what is seen or even what is not, thermionic tubes capable of controlling potent forces under the guidance of less power than a mosquito uses to vibrate his wings, cathode ray tubes rendering visible an occurrence so brief that by comparison a microsecond is a long time, relay combinations which will carry out involved sequences of movements more reliably than any human operator and thousands of times as fast – there are plenty of mechanical aids with which to effect a transformation in scientific records.

Two centuries ago Leibnitz invented a calculating machine which embodied most of the essential features of recent keyboard devices, but it could not then come into use. The economics of the situation were against it: the labor involved in constructing it, before the days of mass production, exceeded the labor to be saved by its use, since all it could accomplish could be duplicated by sufficient use of pencil and paper. Moreover, it would have been subject to frequent breakdown, so that it could not have been depended upon; for at that time and long after, complexity and unreliability were synonymous.

Babbage, even with remarkably generous support for his time, could not produce his great arithmetical machine. His idea was sound enough, but construction and maintenance costs were then too heavy. Had a Pharaoh been given detailed and explicit designs of an automobile, and had he understood them completely, it would have taxed the resources of his kingdom to have fashioned the thousands of parts for a single car, and that car would have broken down on the first trip to Giza.

Machines with interchangeable parts can now be constructed with great economy of effort. In spite of much complexity, they perform reliably. Witness the humble typewriter, or the movie camera, or the automobile. Electrical contacts have ceased to stick when thoroughly understood. Note the automatic telephone exchange, which has hundreds of thousands of such contacts, and yet is reliable. A spider web of metal, sealed in a thin glass container, a wire heated to brilliant glow, in short, the thermionic tube of radio sets, is made by the hundred million, tossed about in packages, plugged into sockets – and it works! Its gossamer parts, the precise location and alignment involved in its construction, would have occupied a master craftsman of the guild for months; now it is built for thirty cents. The world has arrived at an age of cheap complex devices of great reliability; and something is bound to come of it.

A record, if it is to be useful to science, must be continuously extended, it must be stored, and above all it must be consulted. Today we make the record conventionally by writing and photography, followed by printing; but we also record on film, on wax disks, and on magnetic wires. Even if utterly new recording procedures do not appear, these present ones are certainly in the process of modification and extension.

Certainly progress in photography is not going to stop. Faster material and lenses, more automatic cameras, finer-grained sensitive compounds to allow an extension of the minicamera idea, are all imminent. Let us project this trend ahead to a logical, if not inevitable, outcome. The camera hound of the future wears on his forehead a lump a little larger than a walnut. It takes pictures 3 millimeters square, later to be projected or enlarged, which after all involves only a factor of 10 beyond present practice. The lens is of universal focus, down to any distance accommodated by the unaided eye, simply because it is of short focal length. There is a built-in photocell on the walnut such as we now have on at least one camera, which automatically adjusts exposure for a wide range of illumination. There is film in the walnut for a hundred exposures, and the spring for operating its shutter and shifting its film is wound once for all when the film clip is inserted. It produces its result in full color. It may well be stereoscopic, and record with two spaced glass eyes, for striking improvements in stereoscopic technique are just around the corner.

The cord which trips its shutter may reach down a man's sleeve within easy reach of his fingers. A quick squeeze, and the picture is taken. On a pair of ordinary glasses is a square of fine lines near the top of one lens, where it is out of the way of ordinary vision. When an object appears in that square, it is lined up for its picture. As the scientist of the future moves about the laboratory or the field, every time he looks at something worthy of the record, he trips the shutter and in it goes, without even an audible click. Is this all fantastic? The only fantastic thing about it is the idea of making as many pictures as would result from its use.

Will there be dry photography? It is already here in two forms. When Brady made his Civil War pictures, the plate had to be wet at the time of exposure. Now it has to be wet during development instead. In the future perhaps it need not be wetted at all. There have long been films impregnated with diazo dyes which form a picture without development, so that it is already there as soon as the camera has been operated. An exposure to ammonia gas destroys the unexposed dye, and the picture can then be taken out into the light and examined. The process is now slow, but someone may speed it up, and it has no grain difficulties such as now keep photographic researchers busy. Often it would be advantageous to be able to snap the camera and to look at the picture immediately.

Another process now in use is also slow, and more or less clumsy. For fifty years impregnated papers have been used which turn dark at every point where an electrical contact touches them, by reason of the chemical change thus produced in an iodine compound included in the paper. They have been used to make records, for a pointer moving across them can leave a trail behind. If the electrical potential on the pointer is varied as it moves, the line becomes light or dark in accordance with the potential.

This scheme is now used in facsimile transmission. The pointer draws a set of closely spaced lines across the paper one after another. As it moves, its potential is varied in accordance with a varying current received over wires from a distant station, where these variations are produced by a photocell which is similarly scanning a picture. At every instant the darkness of the line being drawn is made equal to the darkness of the point on the picture being observed by the photocell. Thus, when the whole picture has been covered, a replica appears at the receiving end.

A scene itself can be just as well looked over line by line by the photocell in this way as can a photograph of the scene. This whole apparatus constitutes a camera, with the added feature, which can be dispensed with if desired, of making its picture at a distance. It is slow, and the picture is poor in detail. Still, it does give another process of dry photography, in which the picture is finished as soon as it is taken. It would be a brave man who would predict that such a process will always remain clumsy, slow, and faulty in detail. Television equipment today transmits sixteen reasonably good pictures a second, and it involves only two essential differences from the process described above. For one, the record is made by a moving beam of electrons rather than a moving pointer, for the reason that an electron beam can sweep across the picture very rapidly indeed. The other difference involves merely the use of a screen which glows momentarily when the electrons hit, rather than a chemically treated paper or film which is permanently altered. This speed is necessary in television, for motion pictures rather than stills are the object.

Use chemically treated film in place of the glowing screen, allow the apparatus to transmit one picture only rather than a succession, and a rapid camera for dry photography results. The treated film needs to be far faster in action than present examples, but it probably could be. More serious is the objection that this scheme would involve putting the film inside a vacuum chamber, for electron beams behave normally only in such a rarefied environment. This difficulty could be avoided by allowing the electron beam to play on one side of a partition, and by pressing the film against the other side, if this partition were such as to allow the electrons to go through perpendicular to its surface, and to prevent them from spreading out sideways. Such partitions, in crude form, could certainly be constructed, and they will hardly hold up the general development.

Like dry photography, microphotography still has a long way to go. The basic scheme of reducing the size of the record, and examining it by projection rather than directly, has possibilities too great to be ignored. The combination of optical projection and photographic reduction is already producing some results in microfilm for scholarly purposes, and the potentialities are highly suggestive. Today, with microfilm, reductions by a linear factor of 20 can be employed and still produce full clarity when the material is re-enlarged for examination. The limits are set by the graininess of the film, the excellence of the optical system, and the efficiency of the light sources employed. All of these are rapidly improving.

Assume a linear ratio of 100 for future use. Consider film of the same thickness as paper, although thinner film will certainly be usable. Even under these conditions there would be a total factor of 10,000 between the bulk of the ordinary record on books, and its microfilm replica. The *Encyclopedia Britannica* could be reduced to the volume of a matchbox. A library of a million volumes could be compressed into one end of a desk. If the human race has produced since the invention of movable type a total record, in the form of magazines, newspapers, books, tracts, advertising blurbs, correspondence, having a volume corresponding to a billion books, the whole affair, assembled and compressed, could be lugged off in a moving van. Mere compression, of course, is not enough; one needs not only to make and store a record but also be able to consult it, and this aspect of the matter comes later. Even the modern great library is not generally consulted; it is nibbled at by a few.

Compression is important, however, when it comes to costs. The material for the microfilm *Britannica* would cost a nickel, and it could be mailed anywhere for a cent. What would it cost to print a million copies? To print a sheet of newspaper, in a large edition, costs a small fraction of a cent. The entire material of the Britannica in reduced microfilm form would go on a sheet eight and one-half by eleven inches. Once it is available, with the photographic reproduction methods of the future, duplicates in large quantities could probably be turned out for a cent apiece beyond the cost of materials. The preparation of the original copy? That introduces the next aspect of the subject.

To make the record, we now push a pencil or tap a typewriter. Then comes the process of digestion and correction, followed by an intricate process of typesetting, printing, and distribution. To consider the first stage of the procedure, will the author of the future cease writing by hand or typewriter and talk directly to the record? He does so indirectly, by talk ing to a stenographer or a wax cylinder; but the elements are all present if he wishes to have his talk directly produce a typed record. All he needs to do is to take advantage of existing mechanisms and to alter his language. At a recent World Fair a machine called a Voder was shown. A girl stroked its keys and it emitted recognizable speech. No human vocal chords entered into the procedure at any point; the keys simply combined some electrically produced vibrations and passed these on to a loud-speaker. In the Bell Laboratories there is the converse of this machine, called a Vocoder. The loud speaker is replaced by a microphone, which picks up sound. Speak to it, and the corresponding keys move. This may be one element of the postulated system.

The other element is found in the stenotype, that somewhat disconcerting device encountered usually at public meetings. A girl strokes its keys languidly and looks about the room and sometimes at the speaker with a disquieting gaze. From it emerges a typed strip which records in a phonetically simplified language a record of what the speaker is supposed to have said. Later this strip is retyped into ordinary language, for in its nascent form it is intelligible only to the initiated. Combine these two elements, let the Vocoder run the stenotype, and the result is a machine which types when talked to.

Our present languages are not especially adapted to this sort of mechanization, it is true. It is strange that the inventors of universal languages have not seized upon the idea of producing one which better fitted the technique for transmitting and recording speech. Mechanization may yet force the issue, especially in the scientific field; whereupon scientific jargon would become still less intelligible to the layman.

One can now picture a future investigator in his laboratory. His hands are free, and he is not anchored. As he moves about and observes, he photographs and comments. Time is automatically recorded to tie the two records together. If he goes into the field, he may be connected by radio to his recorder. As he ponders over his notes in the evening, he again talks his comments into the record. His typed record, as well as his photographs, may both be in miniature, so that he projects them for examination.

Much needs to occur, however, between the collection of data and observations, the extraction of parallel material from the existing record, and the final insertion of new material into the general body of the common record. For mature thought there is no mechanical substitute. But creative thought and essentially repetitive thought are very different things. For the latter there are, and may be, powerful mechanical aids.

Adding a column of figures is a repetitive thought process, and it was long ago properly relegated to the machine. True, the machine is sometimes controlled by a keyboard, and thought of a sort enters in reading the figures and poking the corresponding keys, but even this is avoidable. Machines have been made which will read typed figures by photocells and then depress the corresponding keys; these are combinations of photocells for scanning the type, electric circuits for sorting the consequent variations, and relay circuits for interpreting the result into the action of solenoids to pull the keys down.

All this complication is needed because of the clumsy way in which we have learned to write figures. If we recorded them positionally, simply by the configuration of a set of dots on a card, the automatic reading mechanism would become comparatively simple. In fact, if the dots are holes, we have the punched-card machine long ago produced by Hollerith for the purposes of the census, and now used throughout business. Some types of complex businesses could hardly operate without these machines.

Adding is only one operation. To perform arithmetical computation involves also subtraction, multiplication, and division, and in addition some method for temporary storage of results, removal from storage for further manipulation, and recording of final results by printing. Machines for these purposes are now of two types: keyboard machines for accounting and the like, manually controlled for the insertion of data, and usually automatically controlled as far as the sequence of operations is concerned; and punched-card machines in which separate operations are usually delegated to a series of machines, and the cards then transferred bodily from one to another. Both forms are very useful; but as far as complex computations are concerned, both are still in embryo.

Rapid electrical counting appeared soon after the physicists found it desirable to count cosmic rays. For their own purposes the physicists promptly constructed thermionic-tube equipment capable of counting electrical impulses at the rate of 100,000 a second. The advanced arithmetical machines of the future will be electrical in nature, and they will perform at 100 times present speeds, or more.

Moreover, they will be far more versatile than present commercial machines, so that they may readily be adapted for a wide variety of operations. They will be controlled by a control card or film, they will select their own data and manipulate it in accordance with the instructions thus inserted, they will perform complex arithmetical computations at exceedingly high speeds, and they will record results in such form as to be readily available for distribution or for later further manipulation. Such machines will have enormous appetites. One of them will take instructions and data from a whole roomful of girls armed with simple keyboard punches, and will deliver sheets of computed results every few minutes. There will always be plenty of things to compute in the detailed affairs of millions of people doing complicated things.

The repetitive processes of thought are not confined, however, to matters of arithmetic and statistics. In fact, every time one combines and records facts in accordance with established logical processes, the creative aspect of thinking is concerned only with the selection of the data and the process to be employed, and the manipulation thereafter is repetitive in nature and hence a fit matter to be relegated to the machines. Not so much has been done along these lines, beyond the bounds of arithmetic, as might be done, primarily because of the economics of the situation. The needs of business, and the extensive market obviously waiting, assured the advent of mass-produced arithmetical machines just as soon as production methods were sufficiently advanced.

With machines for advanced analysis no such situation existed; for there was and is no extensive market; the users of advanced methods of manipulating data are a very small part of the population. There are, however, machines for solving differential equations – and functional and integral equations, for that matter. There are many special machines, such as the harmonic synthesizer which predicts the tides. There will be many more, appearing certainly first in the hands of the scientist and in small numbers.

If scientific reasoning were limited to the logical processes of arithmetic, we should not get far in our understanding of the physical world. One might as well attempt to grasp the game of poker entirely by the use of the mathematics of probability. The abacus, with its beads strung on parallel wires, led the Arabs to positional numeration and the concept of zero many centuries before the rest of the world; and it was a useful tool – so useful that it still exists.

It is a far cry from the abacus to the modern keyboard accounting machine. It will be an equal step to the arithmetical machine of the future. But even this new machine will not take the scientist where he needs to go. Relief must be secured from laborious detailed manipulation of higher mathematics as well, if the users of it are to free their brains for something more than repetitive detailed transformations in accordance with established rules. A mathematician is not a man who can readily manipulate figures; often he cannot. He is not even a man who can readily perform the transformations of equations by the use of calculus. He is primarily an individual who is skilled in the use of symbolic logic on a high plane, and especially he is a man of intuitive judgment in the choice of the manipulative processes he employs.

All else he should be able to turn over to his mechanism, just as confidently as he turns over the propelling of his car to the intricate mechanism under the hood. Only then will mathematics be practically effective in bringing the growing knowledge of atomistics to the useful solution of the advanced problems of chemistry, metallurgy, and biology. For this reason there will come more machines to handle advanced mathematics for the scientist. Some of them will be sufficiently bizarre to suit the most fastidious connoisseur of the present artifacts of civilization.

The scientist, however, is not the only person who manipulates data and examines the world about him by the use of logical processes, although he sometimes preserves this appearance by adopting into the fold anyone who becomes logical, much in the manner in which a British labor leader is elevated to knighthood. Whenever logical processes of thought are employed – that is, whenever thought for a time runs along an accepted groove – there is an opportunity for the machine. Formal logic used to be a keen instrument in the hands of the teacher in his trying of students' souls. It is readily possible to construct a machine which will manipulate premises in accordance with formal logic, simply by the clever use of relay circuits. Put a set of premises into such a device and turn the crank, and it will readily pass out conclusion after conclusion, all in accordance with logical law, and with no more slips than would be expected of a keyboard adding machine.

Logic can become enormously difficult, and it would undoubtedly be well to produce more assurance in its use. The machines for higher analysis have usually been equation solvers. Ideas are beginning to appear for equation transformers, which will rearrange the relationship expressed by an equation in accordance with strict and rather advanced logic. Progress is inhibited by the exceedingly crude way in which mathematicians express their relationships. They employ a symbolism which grew like Topsy and has little consistency; a strange fact in that most logical field.

A new symbolism, probably positional, must apparently precede the reduction of mathematical transformations to machine processes. Then, on beyond the strict logic of the mathematician, lies the application of logic in everyday affairs. We may some day click off arguments on a machine with the same assurance that we now enter sales on a cash register. But the machine of logic will not look like a cash register, even of the streamlined model.

So much for the manipulation of ideas and their insertion into the record. Thus far we seem to be worse off than before – for we can enormously extend the record; yet even in its present bulk we can hardly consult it. This is a much larger matter than merely the extraction of data for the purposes of scientific research; it involves the entire process by which man profits by his inheritance of acquired knowledge. The prime action of use is selection, and here we are halting indeed. There may be millions of fine thoughts, and the account of the experience on which they are based, all encased within stone walls of acceptable architectural form; but if the scholar can get at only one a week by diligent search, his syntheses are not likely to keep up with the current scene.

Selection, in this broad sense, is a stone adze in the hands of a cabinetmaker. Yet, in a narrow sense and in other areas, something has already been done mechanically on selection. The personnel officer of a factory drops a stack of a few thousand employee cards into a selecting machine, sets a code in accordance with an established convention, and produces in a short time a list of all employees who live in Trenton and know Spanish. Even such devices are much too slow when it comes, for example, to matching a set of finger prints with one of five million on file. Selection devices of this sort will soon be speeded up from their present rate of reviewing data at a few hundred a minute. By the use of photocells and microfilm they will survey items at the rate of a thousand a second, and will print out duplicates of those selected.

This process, however, is simple selection: it proceeds by examining in turn every one of a large set of items, and by picking out those which have certain specified characteristics. There is another form of selection best illustrated by the automatic telephone exchange. You dial a number and the machine selects and connects just one of a million possible stations. It does not run over them all. It pays attention only to a class given by a first digit, then only to a subclass of this given by the second digit, and so on; and thus proceeds rapidly and almost unerringly to the selected station. It requires a few seconds to make the selection, although the process could be speeded up if increased speed were economically warranted. If necessary, it could be made extremely fast by substituting thermionic-tube switching for mechanical switching, so that the full selection could be made in one one-hundredth of a second. No one would wish to spend the money necessary to make this change in the telephone system, but the general idea is applicable elsewhere.

Take the prosaic problem of the great department store. Every time a charge sale is made, there are a number of things to be done. The inventory needs to be revised, the salesman needs to be given credit for the sale, the general accounts need an entry, and, most important, the customer needs to be charged. A central records device has been developed in which much of this work is done conveniently. The salesman places on a stand the customer's identification card, his own card, and the card taken from the article sold – all punched cards. When he pulls a lever, contacts are made through the holes, machinery at a central point makes the necessary computations and entries, and the proper receipt is printed for the salesman to pass to the customer.

But there may be ten thousand charge customers doing business with the store, and before the full operation can be completed someone has to select the right card and insert it at the central office. Now rapid selection can slide just the proper card into position in an instant or two, and return it afterward. Another difficulty occurs, however. Someone must read a total on the card, so that the machine can add its computed item to it. Conceivably the cards might be of the dry photography type I have described. Existing totals could then be read by photocell, and the new total entered by an electron beam.

The cards may be in miniature, so that they occupy little space. They must move quickly. They need not be transferred far, but merely into position so that the photocell and recorder can operate on them. Positional dots can enter the data. At the end of the month a machine can readily be made to read these and to print an ordinary bill. With tube selection, in which no mechanical parts are involved in the switches, little time need be occupied in bringing the correct card into use – a second should suffice for the entire operation. The whole record on the card may be made by magnetic dots on a steel sheet if desired, instead of dots to be observed optically, following the scheme by which Poulsen long ago put speech on a magnetic wire. This method has the advantage of simplicity and ease of erasure. By using photography, howcver one can arrange to project the record in enlarged form, and at a distance by using the process common in television equipment.

One can consider rapid selection of this form, and distant projection for other purposes. To be able to key one sheet of a million before an operator in a second or two, with the possibility of then adding notes thereto, is suggestive in many ways. It might even be of use in libraries, but that is another story. At any rate, there are now some interesting combinations possible. One might, for example, speak to a microphone, in the manner described in connection with the speech-controlled typewriter, and thus make his selections. It would certainly beat the usual file clerk.

The real heart of the matter of selection, however, goes deeper than a lag in the adoption of mechanisms by libraries, or a lack of development of devices for their use. Our ineptitude in getting at the record is largely caused by the artificiality of systems of indexing. When data of any sort are placed in storage they are filed alphabetically or numerically, and information is found (when it is) by tracing it down from subclass to subclass. It can be in only one place, unless duplicates are used; one has to have rules as to which path will locate it, and the rules are cumbersome. Having found one item, moreover, one has to emerge from the system and re-enter on a new path.

The human mind does not work that way. It operates by association. With one item in its grasp, it snaps instantly to the next that is suggested by the association of thoughts, in accordance with some intricate web of trails carried by the cells of the brain. It has other characteristics, of course; trails that are not frequently followed are prone to fade, items are not fully permanent, memory is transitory. Yet the speed of action, the intricacy of trails, the detail of mental pictures, is awe-inspiring beyond all else in nature.

Man cannot hope fully to duplicate this mental process artificially, but he certainly ought to be able to learn from it. In minor ways he may even improve, for his records have relative permanency. The first idea, however, to be drawn from the analogy concerns selection. Selection by association, rather than by indexing, may yet be mechanized. One cannot hope thus to equal the speed and flexibility with which the mind follows an associative trail, but it should be possible to beat the mind decisively in regard to the permanence and clarity of the items resurrected from storage.

Consider a future device for individual use, which is a sort of mechanized private file and library. It needs a name, and, to coin one at random, "memex" will do. A memex is a device in which an individual stores all his books, records, and communications, and which is mechanized so that it may be consulted with exceeding speed and flexibility. It is an enlarged intimate supplement to his memory.

It consists of a desk, and while it can presumably be operated from a distance, it is primarily the piece of furniture at which he works. On the top are slanting translucent screens, on which material can be projected for convenient reading. There is a keyboard, and sets of buttons and levers. Otherwise it looks like in ordinary desk.

In one end is the stored material. The matter of bulk is well taken care of by improved microfilm. Only a small part of the interior of the memex is devoted to storage, the rest to mechanism. Yet if the user inserted 5000 pages of material a day it would take him hundreds of years to fill the repository, so he can be profligate and enter material freely.

Most of the memex contents are purchased on microfilm ready for insertion. Books of all sorts, pictures, current periodicals, newspapers, are thus obtained and dropped into place. Business correspondence takes the same path. And there is provision for direct entry. On the top of the memex is a transparent platen. On this are placed longhand notes, photographs, memoranda, all sorts of things. When one is in place, the depression of a lever causes it to be photographed onto the next blank space in a section of the memex film, dry photography being employed.

There is, of course, provision for consultation of the record by the usual scheme of indexing. If the user wishes to consult a certain book, he taps its code on the keyboard, and the title page of the book promptly appears before him, projected onto one of his viewing positions. Frequently-used codes are mnemonic, so that he seldom consults his code book; but when he does, a single tap of a key projects it for his use. Moreover, he has supplemental levers. On deflecting one of these levers to the right he runs through the book before him, each page in turn being projected at a speed which just allows a recognizing glance at each. If he deflects it further to the right, he steps through the book 10 pages at a time; still further at 100 pages at a time. Deflection to the left gives him the same control backwards.

A special button transfers him immediately to the first page of the index. Any given book of his library can thus be called up and consulted with far greater facility than if it were taken from a shelf. As he has several projection positions, he can leave one item in position while he calls up another. He can add marginal notes and comments, taking advantage of one possible type of dry photography, and it could even be arranged so that he can do this by a stylus scheme, such as is now employed in the telautograph seen in railroad waiting rooms, just as though he had the physical page before him.

All this is conventional, except for the projection forward of present-day mechanisms and gadgetry. It affords an immediate step, however, to associative indexing, the basic idea of which is a provision whereby any item may be caused at will to select immediately and automatically another. This is the essential feature of the memex. The process of tying two items together is the important thing.

When the user is building a trail, he names it, inserts the name in his code book, and taps it out on his keyboard. Before him are the two items to be joined, projected onto adjacent viewing positions. At the bottom of each there are a number of blank code spaces, and a pointer is set to indicate one of these on each item. The user taps a single key, and the items are permanently joined. In each code space appears the code word.

Out of view, but also in the code space, is inserted a set of dots for photocell viewing; and on each item these dots by their positions designate the index number of the other item.

Thereafter, at any time, when one of these items is in view, the other can be instantly recalled merely by tapping a button below the corresponding code space. Moreover, when numerous items have been thus joined together to form a trail, they can be reviewed in turn, rapidly or slowly, by deflecting a lever like that used for turning the pages of a book. It is exactly as though the physical items had been gathered together from widely separated sources and bound together to form a new book. It is more than this, for any item can be joined into numerous trails.

The owner of the memex let us say, is interested in the origin and properties of the bow and arrow. Specifically he is studying why the short Turkish bow was apparently superior to the English long bow in the skirmishes of the Crusades. He has dozens of possibly pertinent books and articles in his memex. First he runs through an encyclopedia, finds an interesting but sketchy article, leaves it projected. Next, in a history, he finds another pertinent item, and ties the two together. Thus he goes, building a trail of many items. Occasionally he inserts a comment of his own, either linking it into the main trail or joining it by a side trail to a particular item. When it becomes evident that the elastic properties of available materials had a great deal to do with the bow, he branches off on a side trail which takes him through textbooks on elasticity and tables of physical constants. He inserts a page of longhand analysis of his own. Thus he builds a trail of his interest through the maze of materials available to him.

And his trails do not fade. Several years later, his talk with a friend turns to the queer ways in which a people resist innovations, even of vital interest. He has an example, in the fact that the outranged Europeans still failed to adopt the Turkish bow. In fact he has a trail on it. A touch brings up the code book. Tapping a few keys projects the head of the trail. A lever runs through it at will, stopping at interesting items, going off on side excursions. It is an interesting trail, pertinent to the discussion. So he sets a reproducer in action, photographs the whole trail out, and passes it to his friend for insertion in his own memex, there to be linked into the more general trail.

Wholly new forms of encyclopedias will appear, ready-made with a mesh of associative trails running through them, ready to be dropped into the memex and there amplified. The lawyer has at his touch the associated opinions and decisions of his whole experience, and of the experience of friends and authorities. The patent attorney has on call the millions of issued patents, with familiar trails to every point of his client's interest. The physician, puzzled by a patient's reactions, strikes the trail established in studying an earlier similar case, and runs rapidly through analogous case histories, with side references to the classics for the pertinent anatomy and histology. The chemist, struggling with the synthesis of an organic compound has all the chemical literature before him in his laboratory, with trails following the analogies of compounds, and side trails to their physical and chemical behavior.

The historian, with a vast chronological account of a people, parallels it with a skip trail which stops only on the salient items, and can follow at any time contemporary trails which lead him all over civilization at a particular epoch. There is a new profession of trail blazers, those who find delight in the task of establishing useful trails through the enormous mass of the common record. The inheritance from the master becomes, not only his additions to the world's record, but for his disciples the entire scaffolding by which they were erected.

Thus science may implement the ways in which man produces, stores, and consults the record of the race. It might be striking to outline the instrumentalities of the future more spectacularly, rather than to stick closely to methods and elements now known and undergoing rapid development, as has been done here. Technical difficulties of all sorts have been ignored, certainly, but also ignored are means as yet unknown which may come any day to accelerate technical progress as violently as did the advent of the thermionic tube. In order that the picture may not be too commonplace, by reason of sticking to present-day patterns, it may be well to mention one such possibility, not to prophesy but merely to suggest, for prophecy based on extension of the known has substance, while prophecy founded on the unknown is only a doubly involved guess.

All our steps in creating or absorbing material of the record proceed through one of the senses – the tactile when we touch keys, the oral when we speak or listen, the visual when we read. Is it not possible that some day the path may be established more directly?

We know that when the eye sees, all the consequent information is transmitted to the brain by means of electrical vibrations in the channel of the optic nerve. This is an exact analogy with the electrical vibrations which occur in the cable of a television set: they convey the picture from the photocells which see it to the radio transmitter from which it is broadcast. We know further that if we can approach that cable with the proper instruments, we do not need to touch it; we can pick up those vibrations by electrical induction and thus discover and reproduce the scene which is being transmitted, just as a telephone wire may be tapped for its message.

The impulses which flow in the arm nerves of a typist convey to her fingers the translated information which reaches her eye or ear, in order that the fingers may be caused to strike the proper keys. Might not these currents be intercepted, either in the original form in which information is conveyed to the brain, or in the marvelously metamorphosed form in which they then proceed to the hand?

By bone conduction we already introduce sounds into the nerve channels of the deaf in order that they may hear. Is it not possible that we may learn to introduce them without the present cumbersomeness of first transforming electrical vibrations to mechanical ones, which the human mechanism promptly transforms back to the electrical form? With a couple of electrodes on the skull the encephalograph now produces pen-and-ink traces which bear some relation to the electrical phenomena going on in the brain itself. True, the record is unintelligible, except as it points out certain gross misfunctioning of the cerebral mechanism; but who would now place bounds on where such a thing may lead?

In the outside world, all forms of intelligence, whether of sound or sight, have been reduced to the form of varying currents in an electric circuit in order that they may be transmitted. Inside the human frame exactly the same sort of process occurs. Must we always transform to mechanical movements in order to proceed from one electrical phenomenon to another? It is a suggestive thought, but it hardly warrants prediction without losing touch with reality and immediateness.

Presumably man's spirit should be elevated if he can better review his shady past and analyze more completely and objectively his present problems. He has built a civilization so complex that he needs to mechanize his records more fully if he is to push his experiment to its logical conclusion and not merely become bogged down part way there by overtaxing his limited memory. His excursions may be more enjoyable if he can reacquire the privilege of forgetting the manifold things he does not need to have immediately at hand, with some assurance that he can find them again if they prove important.

The applications of science have built man a well-supplied house, and are teaching him to live healthily therein. They have enabled him to throw masses of people against one another with cruel weapons. They may yet allow him truly to encompass the great record and to grow in the wisdom of race experience. He may perish in conflict before he learns to wield that record for his true good. Yet, in the application of science to the needs and desires of man, it would seem to be a singularly unfortunate stage at which to terminate he process, or to lose hope as to the outcome.

Memex revisited

Vannevar Bush

An Austrian monk, Gregor Mendel, published a paper in 1865 which stated the essential bases of the modern theory of heredity. Thirty years later the paper was read by men who could understand and extend it. But for thirty years Mendel's work was lost because of the crudity with which information is transmitted between men.

This situation is not improving. The summation of human experience is being expanded at a prodigious rate, and the means we use for threading through the consequent maze to the momentarily important items are almost the same as in the days of square-rigged ships. We are being buried in our own product. Tons of printed material are dumped out every week. In this are thoughts, certainly not often as great as Mendel's, but important to our progress. Many of them become lost; many others are repeated over and over and over.

A revolution must be wrought in the ways in which we make, store, and consult the record of accomplishment. This need holds true in science, in the law, in medicine, in economics, and, for that matter, in the broadest subjects of human relations. It is not just a problem for the libraries, although that is important. Rather, the problem is how creative men think, and what can be done to help them think. It is a problem of how the great mass of material shall be handled so that the individual can draw from it what he needs – instantly, correctly, and with utter freedom. Compact storage of desired material and swift selective access to it are the two basic elements of the problem.

I began worrying over this matter more than a quarter century ago, and some twenty years ago published an essay about it, called "As We May Think." Next in this present discussion I want to present some thoughts from that earlier paper. Then we will have a look at what has happened during the past two decades and try to see if we are any closer to the means of the needed revolution.

In that essay I proposed a machine for personal use rather than the enormous computers which serve whole companies. I suggested that it serve a man's daily thoughts directly, fitting in with his normal thought processes, rather than just do chores for him.

If it is to fit in with his normal thought processes, the heart of the matter is selection. Our present ineptitude in getting at the record is largely caused by the artificiality of systems of indexing. When data of any sort are placed in storage, they are filed alphabetically or numerically, and information is found (when it is) by being traced down from subclass to subclass. It can be in only one place, unless duplications are used; one has to have rules as to which path will locate it, and the rules are cumbersome. Having found one item, moreover, one has to emerge from the system, like a dog who has dug up a buried bone, and then re-enter the system on a new path. This is a serious handicap, even with the high-speed machinery just now beginning to be applied to the problem of the libraries.

The human mind does not work that way. It operates by association. With one item in its grasp, it snaps instantly to the next that is suggested by the association of ideas, in accordance with some intricate web of trails carried by the cells of the brain. The mind has other characteristics, of course: trails not frequently followed are apt to fade; few items are fully permanent; memory is transitory. Yet the speed of action, the intricacy of trails, the detail of mental pictures, is awe-inspiring beyond all else in nature.

Man cannot hope fully to duplicate this mental process artificially. But he can certainly learn from it; in minor ways he may even improve on it, for his records have relative permanency. But the prime idea to be learned concerns *selection*. Selection by association, rather than by indexing, may yet be mechanized. Although we cannot hope to equal the speed and flexibility with which the mind follows an associative trail, it should be possible to beat the mind decisively in the permanence and clarity of the items resurrected from storage.

To turn directly to that earlier discussion:

> Consider a future device for individual use, which is a sort of mechanized private file and library. It needs a name. To coin one at random, "memex" will do. A memex is a device in which an individual stores all his books, records, and communications, and which is mechanized so that it may be consulted with exceeding speed and flexibility. It is an enlarged intimate supplement to his memory. What does it consist of?
> *[The author continues with a description of a 'memex' from his previous article. Ed.]*

[...]

Now, is this all a dream? It certainly was, two decades ago. It is still a dream, but one that is now attainable. To create an actual memex will be expensive, and will demand initiative, ingenuity, patience, and engineering skill of the highest order. But it can be done.

It can be done, given enough effort, because of the great advances which have been made in mechanization, the instruments which have already been built in great numbers to aid man's computations and his thoughts, the devices already used for storing and consulting masses of data, the ingenious elements of electric and magnetic circuits which have been developed during the last two decades.

New and powerful instrumentalities have come into use to help it on its way toward birth. Highly sensitive photocells capable of seeing things in a physical sense; magnetic tapes that instantly record with utter faithfulness music or vision; advanced photography which can record not only what is seen but also what is not; transistors capable of controlling potent forces under the guidance of less power than a mosquito uses to vibrate his wings; cathode-ray tubes rendering visible an occurrence so brief that by comparison a microsecond is a long time; transistor combinations which will carry out involved sequences of operations more reliably than any human operator and thousands of times as fast; miniaturization of solid-state devices which will put the complex circuitry of a radio set into the volume of a pinhead; video tapes which put the moving episodes of a football game onto a little strip of film, and instantly reproduce it – there are plenty of mechanical aids with which now to effect a transformation.

So it can be done. Will it be done? Ah, that is another question. The great digital machines of today have had their exciting proliferation because they could vitally aid business, because they could increase profits. The libraries still operate by horse-and-buggy methods, for there is no profit in libraries. Government spends billions on space since it has glamor and hence public appeal. There is no glamor about libraries, and the public do not understand that the welfare of their children depends far more upon effective libraries than it does on the collecting of a bucket of talcum powder from the moon. So it will not be done soon. But eventually it will.

To look forward to memex we will lean on what has already been done. Machines of today fall into two great divisions, first those that supplement man's muscles and his senses, and second those that aid his mind. We do not need to deal with the former, although they have made possible our modern civilization with all its benefits and its dangers. The latter are sometimes included under the general term of thinking machines, but this is an unfortunate expression, for they do not think, they merely aid man to do so. They are of two sorts, analytical machines and data-handling machines, and these are sometimes combined.

The great example of the first sort is the digital machine. It is often called a computer, but this is a misnomer, for the machine does far more than to compute. A single large unit costs several million dollars. Our present business organizations could not operate without it. Properly instructed, it can do about anything a man can do using pencil and paper, and do it a million times as fast. The only things it cannot do are those which distinguish a man from a machine.

It is told what to do by the insertion of a coded tape, and the preparation of this tape is called programming, of which more will be said later. When the computer has completed its job, it delivers its results by rapidly operating a typewriter, or sometimes by drawing them on a screen. It works entirely by using numbers, although these may also represent letters or instructions, and these numbers are in the binary system, that is, to a base two instead of the usual base ten. It gets the numbers it works on from the input tape, or from its own memory, where great masses of data are stored. The tape, and subsidiary instructions stored in the memory, tell it how to manipulate numbers for all its purposes.

Its main element is an elaborate network of electric circuits. These can manipulate numbers by addition, subtraction, multiplication, and division. Thus far it is indeed a computer. But it has, importantly, other circuits which can perform the operations of logic, and it is these which give the digital machine its great power. As a simple example, these can examine a set of numbers and pick out the largest. Or they can follow one set of instructions or another according to the results of the moment. The machine does all of these things very rapidly indeed, many million operations a second.

Another type of analytical machine is the analogue machine. These are nowhere near so precise as the digital machines, but they are far less expensive and are genuinely useful for exploratory purposes, especially in engineering. The principal form is the differential analyzer which appeared some thirty years ago. To use one of these in examining a problem, say the problem of how a suspension bridge will behave in a gusty wind, one assembles an electric circuit which follows the same physical laws as the bridge, though usually with a different time scale, and which then moves a point of light on a screen in just the way in which the bridge will swing in the wind. One has set up an electrical circuit which obeys the same differential equations as the physical system under study, and which hence behaves in the same way, and then one watches it perform, usually by the pattern it produces on an oscillograph.

There are also special-purpose analytical machines which do not belong to either of these classes. An early one of these is the tide-predicting machine. There are also machines for statistical analysis, evaluating correlation coefficients and the like, and for solving integral equations, or interpreting x-ray diffraction patterns of crystals. Some of these have been crowded out by the great success of the digital machine, but they include ideas which should not be forgotten. Everything that can be done on analogue or special machines can also be done on a digital machine, although often not so neatly or flexibly or inexpensively.

Data-handling machines are also of various sorts, from the extremely simple card catalog up through the numerous ways of manipulating punched cards. The memory component of the digital machine is probably the most remarkable of the data-handling devices. Another should be mentioned as well. This is the rapid selector, which first appeared some twenty years ago. This would take a roll of photographic film containing 100,000 or so items in single frames, and select desired items from these in accordance with a code in the margin. It could do so while viewing the items at the rate of 1000 per second. And it printed out the selected items on a short piece of similar film.

Each item could consist of a page of print, drawings, or photographs. There are now a variety of modern forms of this device. Some of them combine the sorting and ordering facility of the punched-card equipment with rapid selection by code. The same sort of thing can of course be done with magnetic tape.

The evolution of data-handling equipment thus has involved two important features: compression, which allows great masses of data to be stored in a small space, and rapid access, by which a single piece of information can be located and reproduced in a very brief time.

The development of detailed devices or elements did not alone make this whole range of equipment possible. There is another, and very important, general consideration which should be noted:

Over three centuries ago Pascal constructed a calculating machine which embodied many of the essential features of recent keyboard devices, but it could not then come into use. The economics of the situation were against it; the labor involved in constructing it, before the days of mass production, exceeded the labor to be saved by use of it, since all it could accomplish could be duplicated by sufficient use of pencil and paper. Moreover, it would have been subject to frequent breakdown, so that it could not have been depended upon; for at that time and long after, complexity and unreliability were synonymous.

Only a century ago, Babbage, even with remarkably generous support, could not produce his great arithmetical machine. His idea was sound enough, but construction and maintenance costs were then too heavy. Inexpensive construction is a new thing. Had a Pharaoh been given detailed and explicit designs of an automobile, and the tools with which to work metal, and had he understood them completely, it would have taxed the resources of his kingdom to fashion the thousands of parts for a single car, and that car would have broken down on the first trip to Giza.

Machines with interchangeable parts can now be constructed with great economy of effort. In spite of much complexity, they perform reliably. It is this reliable complexity, attained at reasonable cost, produced by hard work and the rigors of competition over many years, together with the advance of basic science, and finally man's ingenuity, which has now made it possible to lighten the burden on man's mind, as earlier developments lifted the load from his muscles.

An excellent example of how the advance goes forward is the history of the thermionic tube and the transistor. The thermionic tube was, at its inception, largely a matter of ingenious tinkering, without much reliance on science. Edison, who was no scientist, noted a current from the filament of one of his electric lamps to a plate he put in, but he did nothing about it. De Forest, who probably knew still less science, added a grid between them, and the thermionic tube was born. For many years it was erratic in operation and likely to fail at any moment. Then engineers learned to make a really good vacuum and it became much better. Finally it became so reliable that it could be installed in an amplifier of a submarine cable at the bottom of the sea and expected to last for forty years. It became so rugged that, in the proximity fuzes of the war, it could be put into a shell, fired out of a gun, and still be expected to work as a sensitive electronic detector.

Then came the transistor, which has superseded the tube for most purposes. This certainly did come out of the application of science. A group of men, working on the theory of electric conduction in solids, soon saw how the phenomena they predicted, and checked in the laboratory, could be put to use. The transistor, which can be as small as the eye of a fly, requiring extreme precision of construction, rugged and long lived when once built, is perhaps the most versatile device man has yet produced. With the use of very little power and in a small space it will amplify, modulate, rectify and do dozens of other things. It is one of a whole family of devices based on the use of semiconductors: photoelectric cells, rectifying valves, etc. When the transistor is combined with other elements, resistors and capacitors, sealed in a resin, an assemblage the size of a thimble will do all that used to be done by a radio receiver as large as a suitcase. More than this: by some very modern methods of depositing very thin layers of material in a vacuum, the whole thing can be reduced to a thin wafer the size of a flyspeck, and a thousand such can be produced identically in a single manufacturing operation.

A very great advance – possibly the greatest so far – as we look toward the future memex, is magnetic tape. We have known it for some time in dictating machines. It hit the market modestly soon after the war, and, around 1958, tapes appeared capable of carrying great detail, so much so that video tapes appeared carrying an entire television broadcast with its 70,000 or so complete pictures on a single reel. The idea is a simple one, and the tape is merely a plastic strip covered with magnetic material in finely powdered form. As it passes over an electromagnet, the voice, picked up by a microphone, causes the strength of that magnet to vary, and these variations become impressed on the tape in the form of its magnetization. A wavy form of air vibrations from the voice becomes an identically wavy form of magnetism on the tape. Then, when the tape is run in front of a coil, the voltages there produced can be amplified and fed into a loudspeaker, and a replica of the original voice appears as sound waves in the air.

All this is now ancient history. But it is relatively new to put on the tape the variation of light impinging on a photo-cell as an optical system sweeps its view over a scene, and to do this so that all the details of a complex scene are thus recorded in a small fraction of a second. And then to reverse the process and reproduce the scene to a viewer a thousand miles away.

In our living rooms, we watch a football game. A television camera is scanning the scene line by line, twenty-four pictures a second. The response of its photoelectric equipment, transmitted a thousand miles to our living room, controls the intensity of an electron beam which sweeps over our TV screen and reproduces the play as it occurs. But the output of the camera also sweeps over a fairly broad magnetic tape, and magnetizes it. Thus, a few moments after a play occurs, the tape record can be re-scanned, and the result transmitted to our TV set, so that we can see the play over again. To accomplish this, using a reasonable amount of tape, has required a great compression of the magnetic record. But it has been done and is now accepted by television viewers as a commonplace.

Another important feature of magnetic tape, for our future memex, is that it can be erased. Fortunately, this is easy. One merely sweeps a permanent magnet over the tape and the record is gone. When we take a photograph we are stuck with it; to make a change we must take another whole photograph. But with a magnetic tape which presents to us a picture one can cancel half a line, if he will, add a changed line, or put in a marginal note or code. The moving finger writes, but its record is not here irrevocable.

The advent of the laser may bring photography back into competition for memex storage. It can produce such a small spot of light that there is a factor of 100 or more on compactness compared to magnetic storage. The spot can be intense, so much so that it is used to bore small holes in diamonds, and this means a photographic record can be made in a very short interval, and read out equally rapidly for projection. The film used can be of such low sensitivity that daylight will not affect it appreciably, and the usual processes of development can be avoided, which means parts can be obliterated and additions made to the record.

Beyond this the laser renders possible an exciting process called holography, which may render it possible to project the record so that it is three-dimensional. This is an utterly new form of three-dimensional projection, for it is as though the original scene or model were actually present, and one can move about and view it from various angles. There are many tough problems to be solved before the use of the laser for such purposes becomes practicable. But, for a long view ahead, it exhibits a wholly new field of versatility in which ingenuity will certainly produce results.

There is a point here worth pausing to consider for a moment. For the purposes of memex we need a readily alterable record, and we have it. But alteration of records has a sinister connotation. We watch a girl on the screen moving her mouth and someone else is doing the singing. One can put into a man's mouth for all to hear words he never spoke. The ingenuity which special-effects men use on television is often amusing, sometimes powerfully dramatic, sometimes annoying, as when a razor is seen to shave sandpaper. Advancing technology is making it easy to fool people. It would be well if technology also devoted itself to producing forms of records, photographic, printed, sound-recorded, which cannot be altered without detection, at least to the degree of a dollar bill. But it would be still more effective if the code of morals accepted generally rendered it a universally condemned sin to alter a record without notice that it is being done.

It is thus fairly clear that there is no serious problem today in assembling, editing, and correcting the record, or in compressing it into as small a volume as we may need for memex. If we wish it, a whole private library could be reduced to the volume of a matchbox; similarly, a library of a million volumes could be compressed into one end of a desk. If the human race has produced, since the invention of movable type, a total record in the form of magazines, newspapers, books, tracts, advertising blurbs, correspondence, having a volume corresponding to a billion books, the whole affair, assembled and compressed, could be lugged off in a wheelbarrow.

Compression is important not only to keep us from being swamped, but also when it comes to costs. The material for a microfilm private library might cost a nickel, and it could be mailed anywhere for a few cents. What would it cost to print a million copies? To print a sheet of newspaper, in a large edition, costs a small fraction of a cent. The entire material of a private library in reduced film form would go on ten eight-and-one-half by eleven-inch sheets. Once that was available, with the reproduction methods now available, duplicates in large quantities could probably be turned out for a few cents apiece beyond the cost of materials.

Mere compression, of course, is not enough; one needs not only to make and store a record, to add to it at will, and to erase, but also to consult it. As things are now, even the modern great library is not generally consulted; it is nibbled at by a few. How to consult the new compressed record is a major question in selective analysis.

The great digital computers of today keep their extensive records in various ways. The records constitute their memory, which they consult as they proceed with computation. They use magnetic tapes or disks. But they also use great arrays of minute toroids of magnetic material, interlaced with fine wires. The reason for these latter is the necessity of rapid access. The fast access, in a computer, is fast indeed, and has to be. Times, for them, should be mentioned in nanoseconds, or billionths of a second. In a nanosecond light will move only about one foot. That is why it is important to keep the components of a computer small; its speed of operation is sometimes limited by the time necessary to get an electric pulse from one part to another. The storage in little toroids can respond in times like these.

No problem of speed of access need bother the future memex. Indeed, for memex we need only relatively slow access, as compared to that which the digital machines demand: a tenth of a second to bring forward any item from a vast storage will do nicely. For memex, the problem is not swift access, but selective access. The indication of a possible beginning here is to be found in the rapid selector mentioned earlier. When items on frames projected for viewing can readily have codes entered in their margins, by which they can automatically select other items, we have a significant step toward memex. But the access problem is by no means solved. The storage of memex will be huge, and all parts of it need to be promptly available.

Clearly, we need to study further how the human brain meets this puzzle. Its memory system consists of a three-dimensional array of cells, each cell very small compared to even the volume of magnetic tape used for a single impulse, and the magnetic tape is two-dimensional. We make three-dimensional storage, for example, by an array of toroids, but the units here are huge compared to a cell. Somehow the brain consults this full array and brings into consciousness, not just the state of one cell, but the related content of thousands, to recall to us a scene of a decade ago. We have very little idea as to how it is done. In fact we do not even know what we mean when we write "consciousness." If there is a roadblock in the path toward a useful memex, it lies in this problem of moderately rapid access to really large memory storage.

The heart of this problem, and of the personal machine we have here considered, is the task of selection. And here, in spite of great progress, we are still lame.

Selection, in the broad sense, is still a stone adze in the hands of a cabinetmaker. Yet, in a narrow sense and in other areas, something has already been done mechanically on selection. The personnel officer of a factory drops a stack of a few thousand employee cards into a selecting machine, sets a code in accordance with an established convention, and produces in a short time a list of all employees who are females, live in Trenton, and know Spanish. Even such devices are much too slow when it comes, for example, to matching a set of fingerprints with one of five million on file. Selection devices of this sort have now been speeded up from their previous rate of reviewing data at a few hundred a minute. The great computer will enter its active memory and select a desired item in a micro-second or less, if it is told just where to go for it, and in an interval which is still very brief if it has to hunt for it.

So much for the methods of storing record and of retrieving items from storage. But what about the making of the record? Is it possible that somewhere during this procedure we may find ways of anticipating the selective needs to be encountered later when one wishes to consult that record? Our record-making system of today should remind us of the covered wagon; we are bound to have to improve it, and in doing so we must have an eye to the possibilities of coding, cross-linking, and all else that will be requisite to selective access.

Today, to make a record, we still push a pencil or tap a typewriter. Then comes the business of digestion and correction, followed by an intricate – and largely cockeyed – process of typesetting, printing, and distribution. To consider the first stage of procedure, will the author of the future cease writing by hand or typewriter and talk directly to the record? He does so (indirectly) even now, of course, by talking to a stenographer or into a dictating machine. And there is also the stenotype, that somewhat disconcerting device encountered in court or at public meetings. The operator strokes its keys languidly and looks about the room and sometimes at the speaker with a disquieting gaze. From the machine emerges a typed strip which records, in a phonetically simplified language, what the speaker is supposed to have said. Later this strip is retyped into ordinary language, for in its nascent form it is intelligible only to the initiated. It would be fairly easy to rig a device to operate a stenotype as one talked. In short, if anyone wishes to have his talk directly produce a typed record, all the elements are here. All he needs to do is to take advantage of existing mechanisms – and alter his language.

Our present languages are not well adapted to mechanization. True, digital machines can be made to translate languages, Russian into English, for example. As with their writing of poetry or composition of music, one wonders, not that they do it badly, but that they do it at all. So far, machine translation has not become really useful. But it is improving, and the study that is being devoted to the problem is showing us much about the nature of languages themselves. It is strange that the inventors of universal languages, none of which have ever caught on, have not seized upon the idea of producing one which better fits the technique for transmitting, recording, and modifying speech.

The business of communication between men and machines thus is a complex affair. Men's language has grown without reference to machine use, and now, if we try to talk directly to a machine, it will not understand us. Even if we write or type our material, we have to be careful to put it in form that the machine can grasp.

We see a simple example of this in the numbers put on bank checks with magnetic ink, so that machines can sort them. They have a faint resemblance to figures as we ordinarily write them, but to the machine their altered form is entirely clear.

A better example occurs with the digital computers. These can do extraordinary things, but only if they are given explicit and detailed instructions on how to do them. The process of instruction, programming, uses a special language, incomprehensible to the layman, learned by a human operator only after careful study and experience, but lucid and unambiguous to the machine. There are several new languages under development for this purpose of telling digital computers what to do and how to do it. They are in terms of binary numbers, when they enter the machine, for that is the natural language of the computer.

We will not expect our personal machine of the future, our memex, to do the job of the great computers. But we can expect it to do clever things for us in the handling of the mass of data we insert into it. We particularly expect it to learn from its own experience and to refine its own trails. So our means of communication with it merits careful consideration. Usually we will tell it what to do by pushing a button or moving a lever. Pushing just one button will often call up a fairly complicated internally stored set of instructions. This will serve for ordinary use. But it would be nice, and easily arranged for, if the machine would respond also to simple remarks. If Fido will respond to "lie down," the machine ought to respond readily to such a remark as "hold it."

This matter of a memex learning from its own experience merits some discussion. A digital machine can now be caused thus to learn. Such machines, for example, can be set up to play checkers with a human opponent. Chess is too much for them, because of its complication, which merely means that it calls for an excessive amount of storage and time, but they do very well at checkers. In fact, they can learn to beat a good player. In the digital machine's memory is stored a large number of positions that may occur in a game, and possible following moves to be used. But positions and moves are rated in accordance with assumed values. Confronted with a position, the machine consults its memory and chooses the best-rated move to use. But now comes the real point. It continually alters the rating of the moves in accordance with its success or failure. If a move results in a more highly valued position, its rating goes up, and if it results in catastrophe, it goes down. In this way the machine learns. Playing at first a very poor game, it finally becomes expert.

A memex can be constructed to do similar things. Let's say its master is a mechanical engineer, and that he has a trail which he uses very frequently on the whole subject of heat transfer. The memex notices (we have to use such terms; there are no others) that nearly every time he pursues the trail there are a series of items on which he hardly pauses. It takes them out of the main trail and appends them as a side trail. It also notices that when he comes to a certain item he usually goes off on a side trail, so it proceeds to incorporate this in the main trail.

It can do more than this; it can build trails for its master. Say he suddenly becomes interested in the diffusion of hydrogen through steel at high temperatures, and he has no trail on it. Memex can work when he is not there. So he gives it instructions to search, furnishing the trail codes likely to have pertinent material. All night memex plods on, at ten or more pages a second. Whenever it finds the words "hydrogen" and "diffusion" in the same item, it links that item into a new trail. In the morning its master reviews the new trail, discarding most of the items, and joining the new trail to a pertinent position.

Does this sort of thing sound bizarre or far-fetched? Machines are doing more surprising things than this today.

Much needs to occur between the collection of data and observations, the extraction of parallel material from the existing record, and the final insertion of new material into the general body of the common record. For mature thought there is no mechanical substitute. But creative thought and essentially repetitive thought are very different things. For the latter there are already powerful mechanical aids. We shall need still more.

In particular we have delved far enough into the chemical processes by which the human body operates to grasp the fact that we shall never come to full understanding in this enormously complex field until our processes of reasoning have been greatly refined, and divested of all the clutter of repetitive acts which now take up most of the time that we consider we are devoting to thought. For this reason there will come more machines to handle advanced mathematics and manipulation of data for the scientist. Some of them will be sufficiently bizarre to suit the most fastidious connoisseur of the present artifacts of civilization.

The scientist, however, is not the only person who manipulates data and examines the world about him by the use of logical processes, though he sometimes preserves this appearance by adopting into the fold anyone who becomes logical, much in the manner in which a British labor leader is elevated to knighthood. Whenever logical processes of thought are employed – that is, whenever thought for a time runs along an accepted groove – there is an opportunity for the machine. In fact a machine which will manipulate premises in accordance with formal logic has already been constructed. Put a set of premises into such a device and turn the crank; it will readily pass out conclusion after conclusion, all in accordance with logical law, and with no more slips than would be expected of a keyboard adding machine.

Logic can become enormously difficult, and it would undoubtedly be well to produce more assurance in the use of it. The machines for higher analysis have usually been equation *solvers*. But we now have equation *transformers*, which will rearrange the relationship expressed by an equation in accordance with strict and rather advanced logic. Progress here is a bit inhibited by the exceedingly crude way in which mathematicians express their relationships. They employ a symbolism which grew like Topsy and has little consistency; a strange fact in that most logical field.

What might be the consequences of the developments we have been discussing? Assuredly they would not be limited to the men of science. It could be hoped that the writing of history and biography, for example, would improve, not just in accuracy, but in art, as the writer is able to turn the drudge part of his task over to a tireless assistant, always willing to work when he is, and never at a loss to divine what he wishes to remember. Wholly new forms of encyclopedias will appear, ready-made with a mesh of associative trails running through them, ready to be dropped into the memex and there amplified. The lawyer will have at his touch the associated opinions and decisions of his whole experience, and of the experience of friends and authorities. The patent attorney will have on call the millions of issued patents, with familiar trails to every point of his client's interest. The physician, puzzled by a patient's reactions, will strike the trail established in studying an earlier similar case, and run rapidly through analogous case histories, with side references to the classics for the pertinent anatomy and histology.

Another area in which new machine accomplishments are needed is organic chemistry. These accomplishments are just beginning to appear. There are millions of organic compounds that have been studied, and an unlimited number of possible ones, many of them no doubt useful. The organic chemist is in a tough spot. His memory is severely taxed, and much of his time is consumed in labor that does not call on his true skills. He ought to be able to turn to a machine with a specification of a compound, in terms of either its form or its properties, and have it immediately before him with all that is known about it. Moreover, if he then proposes a chemical manipulation on such a compound, the machine should tell him, within the limits of knowledge at the time, just what will happen. It would do so by using the known laws of chemistry, and the chemist should turn to experiment in the laboratory only for confirmation, or when entering unexplored territory. We are a long way today from such a situation. But machines can certainly do this, if we build them intelligently and then tell them what to do.

The historian, of whom I have spoken above, with his vast chronological account of a people, can parallel this with a skip-trail which stops only on the salient items; he can follow at any time contemporary trails which lead him all over civilization at a particular epoch. There will be a new profession of trailblazers, those who find delight in the task of establishing useful trails through the enormous mass of the common record. The inheritance from the master will become, not only his additions to the world's record, but for his disciples the entire scaffolding by which they were erected. Each generation will receive from its predecessor, not a conglomerate mass of discrete facts and theories, but an interconnected web which covers all that the race has thus far attained.

When the first article on memex was written, the personal machine, the memex, appeared to be far in the future. It still appears to be in the future, but not so far. Great progress, as we have seen, has been made in the last twenty years on all the elements necessary. Storage has been reduced in size, access has become more rapid. Transistors, video tape, television, high-speed electric circuits, have revolutionized the conditions under which we approach the problem. Except for the one factor of better access to large memories, all we need to do is to put the proper elements together – at reasonable expense – and we will have a memex.

Will we soon have a personal machine for our use? Unfortunately not. First we will no doubt see the mechanization of our libraries, and this itself will take years. Then we will see the group machine, specialized, used by many. This will be especially valuable in medicine, in order that those who minister to our ills may do so in the light of the broad experience of their fellows. Finally, a long time from now, I fear, will come the personal machine. It will be delayed in coming principally by costs, and we know that costs will go down, how much and how rapidly none can tell.

It is worth striving for. Adequately equipped with machines which leave him free to use his primary attribute as a human being – the ability to think creatively and wisely, unencumbered by unworthy tasks – man can face an increasingly complex existence with hope, even with confidence.

Presumably man's spirit should be elevated if he can better review his shady past and analyze more completely and objectively his present problems. He has built a civilization so complex that he needs to mechanize his records more fully if he is to push his experiment in its proper paths and not become bogged down when partway home by having overtaxed his limited memory. His excursions may be more enjoyable if he can reacquire the privilege of forgetting the manifold things he does not need to have immediately at hand, with some assurance that he can find them again if they prove important.

The applications of science have built man a well supplied house, and are teaching him to live healthily in it. They have also enabled him to throw masses of people against one another with cruel weapons. They may yet allow him truly to encompass the great record and to grow in the wisdom of race experience. He may perish in conflict before he learns to wield that record for his true good. Yet, in the application of science to the needs and desires of man, this would seem to be a singularly unfortunate stage at which to terminate the process, or to lose hope as to the outcome.

Automata and information

R. A. Fairthorne

In 1713 a lad called Humphrey Potter joined with string the valves of a steam pump so that the pump turned the steam on and off by itself. He had made the first working-class automaton or robot. Till then steam had been used just to push with; you had to turn it on and off yourself, just as you had to tell a horse when to go or stop. Till then, also, automata had been purely aristocratic; just things to look at.

What Humphrey started others have elaborated ever since. The steam-engine looked after itself more and more. Safety valves were put on boilers so that when pressure got dangerously high, the valve would open on its own to let steam off. On faster running engines – pumping-engines were very slow indeed – 'governors' were fitted that cut down the steam supply when the engine ran too fast, and increased the supply when it ran too slowly.

Compared with Humphrey's string and even with the governor, whose behaviour is more profound than elementary textbooks pretend, complicated things like automatic traffic lights, automatic pilots, radar controlled guns, 'electronic brains', and thousands of less-advertised but as useful devices that automatically control machines and processes, have a tendency to appear magical. People ask 'Do calculating machines think?', but these things differ from the strings only in complication. There is a lot more inside them and they can do more things quickly. This may be the sole requirement for psychological activity. If so, unthinking mechanisms, not thinking ones, must be the rarities.

Whether or not self-governing devices think, they have to have the same pattern of behaviour as the simplest governor. In fashionable anthropomorphic terms they must know what is going on that is relevant to their business (and nothing else); they must know what to do about it; and they must be able and willing to do it. Thus the steam-engine governor 'knows' how fast the engine is running by the amount its weights swing out. Nevertheless, the weights will fly out the same amount whether the governor knows about it or not. The governor may or may not know what action to take; it is certainly built so that it cannot take the wrong action, whatever its opinions. Neither reasoning nor self-consciousness is necessary to establish behaviour of the kind required in computers, control mechanisms, and so forth, even if they can learn from experience and even if their actions are unpredictable.

This pattern of behaviour is exhibited most simply in the ordinary governor. The study of such systems is thus called 'cybernetics', after the Greek for 'steersman' from which, through the Latin, comes the English 'governor'. The modern use of the term is due to a group of American mathematicians and neurologists, but the word 'cybernetique' was used in substantially the same sense by Ampère.

In all control systems input organs respond to relevant goings on; the response is signalled to a signal converter that translates them to outgoing signals that start appropriate action in various effective organs. These change the situation which is reported and acted on again, and so it goes on. The signal converter may be called a brain by anyone who finds comfort in so doing; from the point of view of making successful devices it is enough to treat it as a device to alter signals according to rules. The important questions are: What are the limitations to the behaviour of such a device? What rules are needed to get required behaviour that is within these limitations?

In their practical manifestations the rules and signals are of rather overwhelming variety. They may be simple alterations of movements by cams and levers, as when the movement of the arms of a governor produce the required position of the valve. In computing machines the signals are of the telegraphic kind, and represent letters or numbers. The incoming signals are altered so that the outgoing signals correspond to the result of arithmetical operations called for by instructions which may be a part of the incoming signal or a part of the internal state of the control or of both. The internal state of the control can be made so that it depends on what has happened already, and the outgoing signal to depend on this as well as the incoming one and whatever instructions it carries. Thus, the control can alter its own instructions. It is possible, but undesirable, to conceive a whole hierarchy of such converters, each altering the laws for altering the laws for altering the laws, and so on. Mercifully, because the machine is finite, and it works for a finite time, such shoddy Platonism can be avoided by thinking firmly in terms of things and events. In whatever language we care to describe the workings, the basic fact is that the signal transmitted is completely determined by the signal received and the current state of the control. If one has a list of all possible combinations of state and signal, and against each combination the correct response and new state of the machine, the behaviour of the machine is completely determined, provided that it is in working order. Such a list gives no information on how and why such behaviour is produced, and observation of behaviour alone, however complete, cannot answer these questions. For example, in a simple cash register the state of the machine is represented by the total shown on its internal register, the incoming signal represents the price of the item just purchased. The correct response is to signal the price to the display and to change the internal state to represent the sum of this price and the previous total. The internal state does not here affect the outgoing signal, though in more complicated computers it usually does. Whether it does or does not, a list of all possible combinations of total sales and of individual sale values with the correct responses would be adequate but rather bulky; of the order of a million million entries and responses for even the most modest of machines. Evidently in this case, as in all computing, the entries and responses are systematically and simply related, so that the list needed to define or instruct a machine to write out the original list can be very short. It corresponds in fact to any of the sets of rules for correctly adding sterling. The various wheels and cams in the internal register are a particular materialization of these rules for writing out the list of responses.

All such systems of behaviour are logically equivalent to what is called a Thue-system. This is a generalized form of the familiar puzzle in which one word has to be changed to another by altering one letter at a time. However, the rules of substitution may change with the word according to some law, systematic or arbitrary. It seems that the behaviour of all realizable mechanisms is covered by systems of this kind.

What is the implied limitation? No such device can ever deduce the laws of its own operation in complete detail, for this would mean describing itself with itself. This would put it in the position of the map so detailed that it contained a detailed map of itself which therefore contained a detailed map of itself, and so on. Equivalently, it cannot deduce all those and only those true propositions that follow from the laws of its working or, in yet another equivalent, it is not in general possible to predict that one particular word can be derived from another according to the rules of substitution.

Since the laws of operation of the machine are unattainable by the machine itself they must be formulated in advance by the designer, who may be a living creature or a superior machine or, possibly, both or neither. The self-acting machine is therefore the designer's tool, having no more and no less philosophical significance than any other tool. An automaton differs from a screwdriver only in complication, expense, and the fact that the user can give all the guidance needed at one go, and then go off elsewhere, instead of standing over it to guide it all the time. It is true that a chess-playing automaton designed by someone who could formulate the laws for playing or learning to play good chess could play chess better than its designer. A screwdriver drives screws much better than does its designer with his unaided fingers, but this fact, though of equal significance, seems to have caused less stir. All tools do things better than their designers; that is why they are made.

A tool, whether screwdriver, pointed stick, or ACE, is a device for increasing skill; that is, the efficacy of labour. It is not necessarily, or even possibly, a device for reducing the total amount of labour. The tool has to be manufactured from collected and processed materials, and the labour it saves locally in space and time may be at the expense of the future or of others outside the payroll of the originator of the disturbance. The machines that wrap up periodicals tightly, and the use of uninterpreted and uninterpretable alphabetical abbreviations are outstanding examples of devices that do not save but amplify and broadcast labour. Typical of labour delaying devices are photographic records. Without considerable forethought such a mass of data, recorded in so unsuitable a form, may be accumulated that it is more difficult to locate and interpret the record, and more expensive, than to seek the information anew from the original source.

This last illustrates one of the major obstacles to successful use of automata in libraries. The complexity of the machine depends to some extent on the complexity of its rules of operation. If the rules for some particular library activity are formulated, those relating to the essential activity may be fairly to very simple; straightforward multiplication and rearrangement of incoming alphabetical information, for instance. The rules for translating the incoming stimuli into signals suitable for internal processing, and for translating them back into suitable form of output, can be extremely complicated, if known at all. The input stimuli are differing shapes of some but not all marks on sheets of paper of various sizes and other physical characteristics. Interpretation of the marks depends on context (Is this a decimal point, or a flaw in the paper?), and rather widely varying shapes (typefaces, calligraphies, and cacographies) have the same interpretation, which again varies with context. If there is uniform input on punched cards or tape, magnetic tape or wire, and so on, the rules of operation can be confined to the essential selection, marshalling, and automatic transcription, and so forth. That is, the machine rules are thus cut down, but rules for your staff and clients must be multiplied and tightened.

A library takes in as wide a range of information as it can; it does not restrict itself to a particular language, typeface, or format. Thus, current informational automata can be applied with profit only to information specially prepared by libraries for use in libraries. The universal use of typescript does mean, however, that if a machine for reading typescript were commercially available it could very much extend these applications, for it could be coupled not only to automatic typewriters, but also to card and tape punches, thus putting a very large amount of current information within machine range without human transcription. As far too much clerical labour is spent in transcribing the same information into different arrangements, for example, cards to lists and vice versa, and punched tape controlled machines such as the IBM Cardatype can do this, the need for a typescript reader is a very real one.

Ultimately the results of any information equipment must be put out in a form acceptable to human beings. At the very least the information must not be too concentrated in space, time, or symbolism. The automata must be matched to the rest of the processes by suitable information transformers that step down these intensities to intelligible levels. For instance, microfilm demands enlargers, transient displays photographic recording, numerical tabulations graphical presentation. If this is not done, and sometimes it is not, especially in scientific experiment and in computation, the use of high-speed devices creates a state of affairs similar to that arising in a rocking-horse factory when the spots are painted in by hand, but the tails inserted by a high-speed tail-inserter at the rate of several megatails a second.

On the whole, existing automata are not well suited to direct library use.

They have not enough accessible storage, or pseudo-memory; they work too fast; and they demand transcription of their input and, sometimes, output. The scope for any automata in libraries seems to be largely indirect so far as characteristically library activities are concerned. There is, however, need for 'donkey engines' to deal with ancillary work, particularly the residual transport and manipulation left after unnecessary operations have been cut out. Clearly, even the highest-speed machine cannot compete with not having to do the job at all, but there is always a pretty heavy residuum of unavoidable physical labour made up of numerous small and varied duties. What are needed are small robots with good 'muscular memories', less 'intellectual' prowess even than the computing morons, and with strength or dexterity. An existing zoomorph of this type has been made as a camera crane for motion-picture scenes in which separately photographed live action and model action have to be combined. If the scene is to appear as from a moving viewpoint, the movement of the camera over the model must correspond very precisely in space and time with its movements over the live action. To achieve this a history of the movements of the crane wheels and joints and of the film through the camera is recorded automatically on magnetic tape. Later the tape is played back into the crane for the model shot, stimulating servo-mechanisms to move the crane in an appropriate scaled-down version of its earlier activities. Such faithful mimics for transporting, handling, or keyboard activities such as typing the same sequence several times over, have immediate application to libraries, not to say elsewhere. For instance, the amount of transport is usually enough to make mechanical conveyors and handlers worthwhile, if only one route and always that route were used. Powered transport – trolleys and the like – can take any route, but need human control. The proposed device, an 'ape' say, will ape exactly the last operation it has been led through by a human trainer, and will repeat it until modified by any action shown to it once only. Thus it would be as versatile in the sequences of actions it could copy as is the automatic typewriter in the sequences of letters it can copy. An automatic typewriter is, in fact, an ape whose habits are more difficult to change than those of the proposed breed, because they are carried by pre-punched tapes or rolls which cannot be unpunched. One of the greatest virtues of magnetic records is that they are extremely permanent but completely erasable, locally as well as a whole. The values of these sequences of letters or actions depend, as do the products of any tool, on the ability of the users – it is no good saying to an automaton, and untrue saying to a human, 'You know what I mean'. As is well known to all but those who have to find the money for them, good tools enable you to accomplish more for the same effort, not the same with less effort.

Mimicking or echoing automata of the kind specified raise economic problems only. They can be made by connecting magnetic recorders and servo-mechanisms now available to such tools as are appropriate to the class of handling job in mind, from motorized trolleys to typewriter or piano players.

Though automata may not be immediately applicable to a library's informational activities, some of the methods and components used in their design and construction certainly are. As we have seen, their operation depends on translating external stimuli into signals, substituting other signals according to rules that may depend on past history, and translating these new signals into action. Ultimately signals can be made, and therefore information sent, only by inducing changes of power level at some other place and time, preferably that and only that to which you wish to communicate. There are, however, a wide variety of methods by which controlled changes of power level can be produced. Communication engineers are always seeking for observable and controllable phenomena that use as little power as possible. That is why so many eminent physicists and basic physical discoveries come from communication engineering. From the present point of view we can distinguish between two types of remote-controlled power change. In one it is induced by suitably marked or shaped material objects carried from one place or time to another; such are punched cards, books, gramophone records, pictures, and monuments. These are competently named by the old term 'tallies', and their significant marks, shapes, and so on, by 'inscriptions'. As a whole, a tally is therefore a mobile and fairly permanent inscription. At the other extreme, the information may be carried by patterns of transient and immaterial phenomena – pressure, current, voltage, and so on – which are technically 'sign-events' but are adequately covered here by the ordinary implications of the word 'signal'.

For reasonable permanence of information, inscriptions are economically and practically the only medium. Thus, information processing systems that have to hold, rearrange, and store large quantities of information, carry it in the form of tallies. As they work faster, the power demanded increases with the cube of the speed, and a time comes when mechanical handling of material objects becomes impossible, and signals are substituted for tallies.

There are library operations where, for the same reasons though on a different scale, signals are preferable to tallies. Borrowers, especially of technical writings, are not usually interested in the book or document as a material object, but with the information that is or purports to be in it. This information is completely communicated if one communicates, not the document itself, but its appearance. Facsimile signalling can therefore cut out the unavoidable but increasingly overwhelming sequence of booking, packing, dispatching by the lending library and of unpacking, booking, and informing the individual borrower by the borrowing library, and the repeat of these operations when the loan is returned. Moreover, the loan is not available during much of this time, even to the borrower. In a large establishment, or a large and sparsely populated country, the loan period may have expired before the individual borrower receives the book. Television or facsimile telegraph links between libraries would soon save their cost by reducing clerical and handling labour and losses through damage. For small packets of information, television has been used by London banks for some years, and more recently for drawings and diagrams by airfield controls. The extra item needed for library application is a practical office device in which sensitized paper is put in contact with the image tube and developed immediately and locally to give a reasonably legible version of the transmitted image. The paper could be made impermanent to avoid legal difficulties.

In short, modern technology has extended the range over which we can read, and has made it possible for libraries to change over from transport to display in some of its activities. What modern technology offers, modern finances may deny, but the trend has started.

For some time the main benefit of automata to libraries will come from application of the principles of their construction rather than from actual automata. To construct one, we must first know what we want it to do; a task of some difficulty because the specification must be precise. Knowing what it is to do, we must then formulate the rules for doing it or learning to do it. Usually there are advantages in building automata ready trained, but some that learn by experience have been made for practical jobs. Certain experimental telephone exchanges save the cost of much capital equipment by adapting themselves to subscribers' habits. An ordinary exchange is built on the assumption that any subscriber is equally likely to call any other. These experimental exchanges keep a count of those numbers most frequently called by each subscriber and adjust their workings accordingly.

The need to adapt methods to customers' habits is not a novel discovery so far as libraries are concerned. Normally it is done by human beings according to rules that they themselves cannot formulate, and perhaps would not be worth mechanizing if they could. This may be true also of other library activities supposed to be essentially human. Even if there is no idea of building an actual machine, attempts to formulate its rules of operation may be surprisingly profitable. As we have seen, tools can only increase the skill with which operations are carried out. Thus, their guiding rules can be tried out by human beings carrying out pen and pencil operations on symbols strictly according to rule. The rules for translating data into machine form, which we have seen to be very complicated in general, can be omitted. What remain are the rules logically equivalent to the basic activity. After these have been tried out by hand, three questions may arise: Even if the automaton is not built, can these rules be used as a routine by human operators? Is it profitable to build a machine for the basic task with or without the data served up in palatable form by human transcribers? If we had a machine with all our own experience and information relevant to this task built in, what is the best it could do?

For one seemingly intellectual activity for which machine rules have been formulated, the formation has answered the first two questions affirmatively. This is translation of mathematical German texts into English. Such texts have sharply defined meaning, and the meaning of one sentence is largely independent of the meanings of its neighbours. The rules have been made to suit a particular automaton, the U.S. Bureau of Standards West Coast Computer, SWAC. There are therefore precise bounds to the size of vocabulary and the amount of context that can be considered at one time.

This specific task has shown that the difficulties of translation are distributed otherwise than was expected. The main task is not to deal with syntax and idiom, but to construct the vocabulary. Syntax is dealt with by generalizing the concept of 'part of speech'. To each word of the original text is ascribed by rule a number depending on its immediate neighbours as well as on itself. When a sentence or self-contained clause (determined automatically) is rearranged in order of these numbers, the words will be in the order appropriate to English. The real difficulty lies in the word-for-word translation remaining. The most efficient dictionary for machine use is one in which each item is, on the average, referred to equally often. Otherwise some will be passengers taking up valuable space on a magnetic drum. Ordinary dictionaries are not like this, for they either attempt to list all words in use, or are for use by people who have some sort of mental dictionary fitted already. The machine dictionary must be consulted for every word. This, and the equal load condition, implies amongst other things that it must contain the right proportions of the different parts of speech. Evidently once this dictionary has been compiled, it and the syntactical numbering rules could be used for routine translation by people with no knowledge of German.

Similar analysis of other activities for automatic execution may prove profitable in the same way, even if the ironmongery is not made.

The 'wonderful one-horse shay' principle of equally distributed average use is of fundamental importance. It is basic in what is rather misleadingly named information theory – to wit, the theory of signals – but its applications to library work lie outside the scope of this paper.

The translation project demonstrates also that the problem of context can be dealt with by routine operations. It has always been a worry, especially in library classification, that the semantic content of words depends on the pragmatic context of their production. They are, in Peirce's terms, 'indexes', not 'symbols' or 'ikons'. Another independent attack on another aspect of context has been launched by Richens. The same basic principle lurks also behind Shannon's work on the prediction of English from its context. We need not wait to drag it completely into the open before attempting an empirical solution to an urgent problem, the retrieval of information on experimental techniques. This is a topic inaccessible to those parts of library classifications that tie themselves too closely to, even if they do not confuse themselves with, the classification of knowledge and of phenomena.

Shannon's work referred to above shows the need for combined experiment and theory in the informational field. Moreover, experiment that does not need grey-enamelled and chromium-plated apparatus. He wished to know, as an ideal standard of performance for communication coding devices, the maximum compression of English text that could be achieved by a machine having an adult's experience of the spelling and structure of printed English. This is clearly less than one-half, because if one-half of either the words or of the letters of a printed page are obliterated at random, the text is still intelligible. Beginning with ingenious experiments on willing victims to find the chance of guessing a letter correctly when the hundred preceding or following letters have been given (the odds turn out to be about even), he imagined transmitters and receivers that were, in linguistic experience, identical twins of the average human experimenter. From this he was able to specify the list of stimuli and responses necessary to determine the logically equivalent automaton, and to estimate its performance quantitatively. The list has to include all possible English sequences of a hundred letters, spaces, and points. It is not quoted in full. These experiments and their analysis show that English text can be compressed to rather less than a quarter without losing information, a point of some importance in spelling reform, and in the design of shorthands that can be read by others, rather than those that are jogs to the writer's memory. It can also shorten somewhat the labours of the famous monkeys with their typewriters.

Compressed text is necessarily at the mercy of transcription errors. As in a sequence of numerals, each symbol will be independent of the others. Thus, it is now not only impossible to correct errors, but even to detect them, from the context.

Here again it is the work behind automata, rather than the automata themselves, that has value to libraries. On the whole the greatest contribution cybernetics will make to information services will be to lay bare by theory and experiment, in which libraries can share, principles and phenomena that are common to both. Sensational advances, suggested by diluted analogies between control mechanisms and multi-purposed institutions, are not to be expected. The movement of the apple had to be studied before that of the moon, but we have reached a stage where theory can be applied to small, clear-cut, practical problems with some chance of success.

Economics of inquiring, communicating, deciding

Jacob Marschak

We hear much of today's "informational revolution." We are also told of the rapid growth of the "knowledge industry." Informational revolution is exemplified by TV pictures of the moon surface and also by robotized stock market transactions and, hopefully, by computerized professors. Fritz Machlup defined the knowledge industry to include education and research as well as publishing and broadcasting. He estimated its share in the gross national product of 1958 at 23 per cent to 29 per cent, and its growth rate at about 10 per cent, or twice that of the GNP. Projecting to the present, the share of the knowledge industry would then appear to straddle the 40 per cent mark!

There is a suspicious overlap between these activities and those which Adam Smith and Karl Marx called "unproductive" and which include the work of kings and professors, none of whom add to the vendible and visible stocks of the nation. To be sure, recent analysis – for example, by T. W. Schultz and Carl V. Weizsaecker – found it both convenient and feasible to define human capital and thus to consider education as investment. But the notable fact remains that professors and kings or the modern equivalent of kings – managers, both public and private – are strongly involved in those trends: informational revolution and growing knowledge industry.

Not only professors and managers, but also computers and TV sets, are involved in still another trend relevant to my talk. A growing proportion of both man-hours and machine-hours is not employed for using large amounts of energy (muscular or otherwise) to transform or transport matter. Instead, so-called "brains" (human or otherwise) are employed to manipulate symbols. A sequence or network of such symbol manipulators uses up a minute amount of energy to eventually release, trigger-like, large amounts of energy through the more brutal medium of generators, muscles, and machine tools. In a modern assembly or disassembly plant (sawmill, meat packing), a growing majority of people, wearing white collars, or blue denims as well, do the brain work of inspecting, deciding, reporting – shunting, pushing buttons – and not the muscular work of shaping or carrying material masses; and a growing proportion of machines, called control mechanisms, are also busy with inspecting, reporting, deciding, and not with transforming or transporting matter and energy.

My topic is the economics of what I shall call the services of inquiring, communicating, deciding. Data are gathered. They are communicated to the decision-maker. He, on the basis of the message received, decides upon the action to be taken. A higher-order decision must have been made previously. Someone representing the interests of the economic unit considered – its head, leader, organizer – must have chosen a particular combination of these three services from all those available in their respective markets. The maker of this higher-order decision (the "meta-decider," sometimes called "the organizer") may happen to be the same person who will decide upon acting. Or more generally, the organizer will hire the services of the decision-maker – who, in appropriate cases, may be just a robot.

I might also call my topic the economics of the instruments, or devices, human or otherwise, for inquiring, communicating, and deciding. For it is not relevant, for my purposes, to distinguish between purchased instruments and hired services, provided the length of the hire contract is specified. In any case, I shall be concerned with symbol manipulators, human or otherwise, rather than with processors or transporters of matter or energy.

Here is what I plan to do. I shall present, in turn, from the user's point of view, the successive links in the sequence of symbol-manipulating services: inquiry, or data gathering; communication of messages; and deciding upon actions on the basis of messages received. It will turn out, in fact, that the link called "communication" must be broken into two distinct services: on the one hand, the service of "encoding and decoding" which, at the present state of arts and in the most numerous and socially most important cases, is best supplied by men; and on the other hand, the service of "transmission" which is best supplied by inanimate equipment. As to the supply conditions of services of inquiry, or data production, and of decision making, I shall be able to submit nothing but crude illustrations, I am afraid. As to the demand side, economists will not be surprised that to make an economical – that is, optimal, efficient – choice the user must choose those links, or components, simultaneously (just as a manufacturer cannot choose between rail and road as means of bringing him fuel without making up his mind, at the same time, whether the fuel should be coal or oil). Hence, the jointness of demand for services of inquiry, communication, and decision.

To be sure, current engineering science finds it convenient to isolate a pure theory of communication – a theory of efficient coding and transmission alone, essentially created by Claude Shannon and streamlined by Jack Wolfowitz. At the other extreme, statistical decision theory culminating in the work of David Blackwell leaves out the communication component and only analyzes, from the point of view of a perfect decision-maker, the optimal choice of inquiry, or data producing, services, also called "experiments." I shall later state the implicit tacit assumptions made in each case. If they are not satisfied, the user guided by those subtheories will have suboptimized. This is not to say that we ought not to break up a complex problem into subproblems, assuming them independent as a provisional first approximation. Given our human limitations, this may even be the optimal research strategy. It just happens that the economist is aware of interdependencies: he calls them complementarity and substitutability of goods. He is also traditionally permitted – as is the philosopher – to attack complexities with ridiculously simple examples in order to get directly to the general and fundamental.

Let me, then, go ahead with a simple example. I must decide this Thursday night whether to fly West next Saturday morning. Visibility and winds along the airplane's route the day after tomorrow will determine whether, if I do fly, I shall arrive in time for an important appointment or shall be killed in a crash. If I don't fly, I miss the appointment. But I cannot know what the weather will be. Instead, I may look tonight at the hotel barometer; or I may rely on the radio reports of other, more numerous and accurate barometer readings; or I may rely on the *Farmer's Almanac*. If the cost of these various services were equal, I would choose the one which gives data most closely reflecting (in some sense) the actual event I am interested in: the weather on Saturday. But perfection is costly, and I shall choose a service that is known not to mislead too grossly or too frequently, yet will be relatively cheap. Take another example. A store's profit will be influenced by its inventory policy, given the actual future demand for its merchandise. Lacking the knowledge of this demand, the firm will have to choose between various services of market forecasters differing in precision and accuracy but also in the fees charged.

So much about services that inquire; i.e., produce data. These data are not identical with, yet do reflect in some sense the events that are relevant to, the result of a decision. Now, the decision-maker may or may not be able to obtain such data directly. Another service called "communication" will bring to him, not those data, but a message, possibly delayed or distorted, about those data. He must decide on the basis of such a message, which is now twice removed, in a sense, from the actual, result-relevant event: weather on Saturday, demand next month, and so on. The inventory example illustrates also the nature of decision services. Inventory policy is a rule stating whether and how much to reorder when the stock at hand is at a given level and you have some – usually imperfect – knowledge related to the prospective demand of your customers. One policy is similar to the one you use when you decide whether to refill your car's oil tank up to a certain upper level (this you will do whenever oil is below a certain lower level) or to leave it alone. Except that in the inventory case the two critical levels themselves are not fixed but should depend on what the store has learned – however imperfectly – about future demand; that is, on the message it has received about the data produced by a market forecast. Such a decision rule or strategy – a rule of how to respond to the message – may require the sophisticated services of a specialist or a computer. Contrast with this a simple routine rule: to refill the inventory every Monday to a constant level. This can be handled by an unskilled clerk or a cheap robot. The more sophisticated, flexible, nonroutine rule would be preferable if it were not for its higher cost.

To state more precisely the problem facing the user of the data producing, communication, and decision services, it is convenient to represent each service as a transformer of inputs into outputs. (Transformer, transformation, and function mean essentially the same thing.) A data producing service such as a barometer is a transformer whose input is the result-relevant event (weather next Saturday) and whose output is an observed value, a datum (the barometer reading tonight). We say that the data service is both precise and reliable if to each result-relevant event corresponds one observation or datum, and conversely. But this perfection is almost never attained. Generally, each event may be reflected in various alternative observed values, with some alternatives more likely than others. We have here the case of an "unreliable" (probabilistic, stochastic, noisy) transformer. For example, suppose that if Saturday's weather is going to be good, the chance that the barometer shows high pressure tonight is 80 per cent. We say that the likelihood of the observation "high pressure," given the event "good weather," is 80 per cent. Suppose the likelihood of low pressure if the weather is going to be dangerous is also 80 per cent. Suppose that on a second barometer both these likelihoods are lower: 60 per cent and 60 per cent, say. If you have access to both barometers at the same cost or effort, you will prefer to be guided by the first one. For, in an obvious sense it is more reliable (more informative, in Blackwell's terminology). Indeed, in the case of perfect reliability the two likelihoods would be 100 per cent and 100 per cent; and clearly our first barometer (with 80 per cent, 80 per cent) comes closer to this perfection than the second (with 60 per cent, 60 per cent). In fact, the second comes closer than the first to the other extreme: likelihoods 50 per cent, 50 per cent, in which case the barometer would be useless.

Or consider a consumer survey conducted by a government agency. The success of the government decision to undertake one or another administrative action depends on the attitudes of all consumers. But only a certain number are sampled. The larger the sample the more reliable is the estimate of people's attitude. But also the more expensive. It is not different with research into the laws of physics or biology. What I called "result-relevant events" the statisticians call "hypotheses." The data producing service (e.g., a sampling) they call "experiment," and the data are called "observations."

Unfortunately, it is not always possible to compare two data producing services on the basis of the likelihoods only. How does our first barometer, with the 80 per cent likelihood of high pressure given good weather, and 80 per cent likelihood of low pressure given dangerous weather compare with the following one: if the weather is going to be good, the barometer will show high pressure for sure, i.e., with likelihood 100 per cent; but given dangerous weather, it will show high or low pressure, not with 80:20 but with 50:50 chances. Thus, whenever the new barometer shows low pressure, it gives you absolute certainty that the weather cannot be good. But when it shows high pressure you are left guessing, and you might be better off with the original barometer. Which barometer should guide you?

Here I must remind you that, just as economists are Keynesians or non-Keynesians, the statisticians are Bayesians or non-Bayesians. The Bayesians, having given serious thought to our problem, tell me to consult two further items: (1) the approximate, so-called "prior," probabilities that I would assign to dangerous versus good weather for next Saturday – in the absence of any barometer, e.g., on the basis of my previous experience with December weather; (2) the utilities that I assign to various results of my actions and whose actuarial value I would like to be as high as possible. In our case, the best result is: surviving and making the appointment. The second best is: surviving but missing the appointment. The worst is death. What matters is the ratio of the disadvantage of missing the appointment (but staying alive) to the disadvantage of death. Now, unless a barometer is useless, I would fly if it shows high pressure and not fly otherwise (or vice versa). The probabilities of these pressure readings depend on my prior probability of the two states of weather and on the likelihoods characterizing each of the two barometers. Therefore, those pressures will be read on the two barometers with different probabilities. Hence the average utility of the results of actions dictated by each barometer and weighted by the probabilities of its readings will differ as between the two barometers. I prefer the one whose dictation will yield, on the average, the higher utility.

Suppose, then, the prior probability of bad (i.e., deadly dangerous) weather is, in my judgment, about 10 per cent. Should I use the "old" barometer (whose likelihoods are 80 per cent and 80 per cent) or the "new" barometer (whose likelihoods are 100 per cent and 50 per cent), still assuming that the costs are equal. It turns out that I should stick to the old barometer as long as I judge the disadvantage of missing the appointment (while staying alive) to be less than one-seventh of the disadvantage of death. This is surely my case!

I believe this business of assigning prior probabilities to events and of utilities to results is a headache familiar to cost-benefit analysts, certainly present in this very room! It surely requires some soul-searching to appraise and reappraise the subjective probabilities ("beliefs") and utilities ("goals," "values," "tastes") of your government agency. You will presumably try out various plausible assumptions and see whether your boss likes the decisions derived from them; and whether under repeated trials of this kind his choices reveal a consistency of his beliefs and of his tastes. Or perhaps such trials will gradually train him toward consistency – toward learning what he wants and believes.

So far, we have seen how the consistent user should choose between available inquiry services when their costs are equal. This has required, for some though not all pairs of such services, to take account of the user's utilities and prior probabilities in order to compute the average utility attainable on the basis of an inquiry. If costs are not equal, the knowledge of prior probabilities and utilities becomes necessary in all cases. For simplicity, a tacit assumption is made. One assumes, in terms of our example, that it is possible to represent the utility of, say, "having made the appointment and having my wealth diminished by the dollar cost of a particular inquiry" as a sum of two utilities. In other words, utility is assumed to be additive and to be commensurable with dollars. Under this assumption, one may call the average utility of results of actions guided by an inquiry simply its dollar value to the user, his demand price. He compares it with the dollar cost, which is the supply price. The additivity assumption is implicitly made by statisticians. They assume in effect that the disutility of a result of action based on a sampling survey is measured by the size of the error of an estimate, and then vaguely compare its average with the dollar cost of the sampling survey. Engineers indulge in similar practices, as we shall see. Not so the economists. Sophisticated in matters of substitution, income elasticity, and risk aversion – all of which question the assumption of additive utility – they raise a warning finger. They do so at least when they talk theory and are not themselves engaged in the practical pursuits variously called "management science," "operations research," "system analysis," "cost-benefit analysis."

My barometer case – with just two alternative events, two possible observations, and two actions – is probably the simplest nontrivial problem in statistical decision theory. As I said before, this theory neglects the communication link. Result-relevant events, each having some prior probability, are transformed into data by a transformer called an "inquiry," or an "experiment" (like a barometer, or a sampling survey). Data flow directly into the transformer called "decision-maker," who applies a decision rule (e.g., "fly if barometric pressure high") and puts out actions. Finally, events and decisions are joint inputs of a transformer which may be called "result function": its output is a result. Assuming additive utility, a dollar amount can be attached to each result. And the probability of each result is determined by the prior probability of each event, by the array (characteristic of the data producing service) of the likelihoods of the data, given each event, and by the decision rule (characteristic of the given decision service) transforming data into actions. The probabilities of the results thus derived serve as weights to obtain the average of their utilities. This average may be called the "gross value," to a given user, of the given pair of data producing and decision-making services. It is the maximum demand price offered by their user. Their combined cost asked by the suppliers is the minimum supply price. The difference may be called "the net combined value of these two services to the user." He will choose the combination with highest net value.

This net value depends, then, on the one hand, on the choice of services made by their user. On the other hand, it depends on conditions outside of his control: viz., his utilities and prior probabilities and the costs of available services. His problem is to maximize the net value of the data producing and decision-making services, given those noncontrolled conditions.

Those familiar with what has been called "information theory" in the last two decades will have noticed that, so far, we have not used the concept of amount of information, measured in units called "bits." My uncertainty about a set of alternative events is the same as the amount of information that I would receive if that uncertainty were completely removed; that is, if I would know exactly which particular event does happen. Roughly speaking, this uncertainty is measured by the smallest average number of successive yes-and-no answers needed to completely remove uncertainty. This number depends roughly on the prior probabilities of events. Suppose, for example, that the following four events have equal probabilities (one-quarter each): the bride at the neighborhood church next Saturday will or won't wear a mini-dress, and her dress will or won't be white. To learn the color of the dress I need one yes-or no question; so my uncertainty about color measures one bit. For the same reason, my uncertainty about both color and style is two bits since I need two yes-and-no answers to answer it. Thus the uncertainty, measured in bits, about those two mutually independent sets of events is the sum $(1 + 1 = 2)$ of the uncertainties about each of them. The number of bits is, in this sense, additive: a property that we require of every genuine measure, such as that of time, distance, volume, energy, dollar income, and dollar wealth.

If the four bridal events were not equally probable – for example, if the odds for a maxi-dress were not 1:1 but 9:1 (while a dress of each style were still as likely as not to be white) – the average necessary number of yes-and-no questions and thus the number of information bits would be smaller in the long run, i.e., over a long sequence of such events: for we can then profitably identify the more probable sequences (i.e., those mostly consisting of maxi-dresses) by asking a few questions only – as any skilled player of the "20 questions game" knows. As before, the count of bits agrees remarkably with the intuitive use of the English word uncertainty: for when the odds are 9:1 I am almost certain, and with odds 1:1, I am fully ignorant, am I not?

Now suppose you have the choice between learning both the style and the color of the bride's dress and learning, with equal speed and for the same fee, the future price of your stocks. Suppose the price is as likely to rise as to fall. Depending on your selling or purchasing now, you may lose or double your fortune. A service that will tell you correctly whether the stock price will rise or fall conveys only one bit of information; whereas the service telling you correctly both the style and the color of the dress provides two bits. Yet you will prefer to learn about the stocks, not about the dress. There is, thus, no relation between the number of bits conveyed and the gross value of the data producing service. Nor does there seem to be a relation between the number of bits and the cost of a data producing service. For example, the cost of a sampling survey depends on its size, and this is not clearly related to the number of bits involved.

On the other hand, the number of yes-and-no symbols involved is clearly relevant to the performance and the cost of the transmission service regardless of whether these symbols refer to the length of the bridal skirt or to the trend of prices of your stock. To the economist, the contrast between production and transmission of data is strikingly analogous to the contrast between production and transportation of goods. A gallon of whiskey is more valuable than a gallon of gasoline: their costs to the producer and their values to the buyer are quite different. Yet to transport one gallon over one mile costs about the same for any liquid.

When, some twenty years ago, those elusive things labeled by the vague English words "uncertainty" and its negative, "information," were harnessed, subjected to genuine measurements (as was energy some hundred years ago, and mass and force much earlier), it was easy to understand the enthusiasm of people in elusive fields such as psychology. But also, to some extent, in statistics and in mathematics, where it was partly due to deep and beautiful theorems developed in this context. It is remarkable that C. Shannon who first proposed these theorems clearly limited their application to communications.

Statistical decision theory deals only with the choice of experiments and of decision rules; that is, with the choice of data producing and of deciding services. It omits the lower row of the chart reproduced above: encoding of data into signals, transmitting signals through a "communication channel," and decoding them back into messages that the decider would understand. In other words, for the statistician the decision is taken on the basis of a message which is simply the same thing as the data produced by the inquiry or experiment.

Not so with the communication engineer. His responsibility is to construct channels for the fast and reliable transmission of signals. (It all started in the telephone industry!) He is therefore also interested in devising appropriate codes which translate ordinary English into signals and signals back into English. But, to concentrate his mind on pure communication economics he makes, in effect, the following simplifying assumptions: First, events and data are identical, for he is not interested in the imperfections of the data producing service. Second, deciding is the same thing as decoding; so that action is simply the same thing as the message received. Third, as we have observed for the case of non-equiprobable events (which is, of course the usual case), the count of bits presupposes, in general, long sequences of events; and, as we shall see, such long sequences are also essential to make the crucial concept of "channel capacity" useful. Fourth, in most though not all of their work, communication engineers assume an extremely simple result function. There are only two results: bad (say, minus one), when the decoded, received message is not identical with the datum sent; good (say, zero), when the two are identical. That is, all errors are equally important, have the same disutility, whether an inch is taken for a mile or merely a colon is taken for a semicolon. Finally, utility is assumed to be additive; i.e., it is conceived as the sum of certain measurable advantages and disadvantages, appropriately converted into dollars. We have seen that statisticans make the same assumption when they compare the sampling error with the dollar cost of the sample. The economist who detects and warns against this assumption is somewhat of a purist. The assumption is surely convenient for practical purposes and its removal is perhaps not that urgent.

Indeed this last assumption permits the engineer to ask the following economic question on behalf of the user of transmission channels and of coding services. Given the dollar costs of available channels, what is the best combination of the following evils: the probability of error (any error); the cost of the channel; the average time delay, which depends both on the length of signal sequences transmitted at a time and on the size of the code's vocabulary. That is, disutility is thought of, in effect, as a sum of dollars that buy a given channel; plus the dollar-equivalent of an error (any error); plus the dollar-equivalent, to a given user, of each time delay arising in the coding and transmission of a given datum. The user's problem is to choose that combination of channel and code which will minimize the sum of the averages of these amounts, weighted by appropriate probabilities. What do these probabilities depend on? On the uncertainty about data (= events); on the likelihood array characterizing the channel's reliability (the array of conditional probabilities of output symbols given an input symbol); and on the coding and decoding procedures.

Clearly, an appropriately redundant code can almost overcome the lack of reliability of the channel; that is, it can almost eliminate the occurrence of errors. For example, the encoder just lets every "yes" or "no" be repeated many times, and the decoder takes "no" for the answer if he has received more "noes" than "yeses". "Don't! – repeat, don't! – repeat, don't shoot!" If I have heard two don'ts and only one do, I shan't shoot. However, we may need great redundancy of the code if the channel is very unreliable; and this will cause long delays if the channel is slow. But a channel that is fast and reliable is expensive.

If the user can afford to wait for a long sequence of data to flow in before they are encoded, the problem of choosing between channels is simplified, for their variety is reduced as follows. Instead of a whole array of likelihoods (of channel output symbols, given each input symbol) it becomes sufficient to use a single reliability measure (in bits per input symbol) which, multiplied by the channel's speed of transmission (in input symbols per time unit), gives the channel's "capacity," in bits per time unit. Provided this capacity is larger than the number of bits per time unit that characterizes the uncertainty and speed of the flow of data, it has been shown that the user can achieve any desired probability of errors, however small, by using an appropriate, though redundant, code. Assuming that such codes have indeed been constructed (quite a difficult problem, solved only to a small part), it would be for the user to weigh against each other the disadvantages of errors, of time delays and of the high costs of high-capacity channels.

To avoid errors in our mutual understanding, let me be redundant, mindful of my low transmitting capacity and of your limited memory. I said a short time ago that engineers have isolated the pure communication problem by not concerning themselves with the services that produce data and that decide on acting; and also by usually refusing to distinguish between important and unimportant errors. I also pleaded, a longer time ago, on behalf of economists who emphasize that the demand for all services, all the transformers on my chart, is a joint one. Indeed, the user can improve the reliability of messages on which decisions are based by improving the communication service, but also by improving the data producing service which he is free to choose. Similarly, the user (the "meta-decider") is free to choose the deciding service; for example, he may prefer not to burden the unskilled but inexpensive decider with messages written in a vocabulary that is too rich and fine. Moreover, depending on the user's result function, he may fear some errors of communication but be indifferent to others. He may be indifferent to the music of the voice at the other end of the telephone; so he does not really need a high-fidelity telephone.

On the other hand, statisticians have isolated their problem, also essentially an economic one, by omitting the communication components. As I said before, this may be a good research strategy. I am told that in the early space vehicles rectangular pieces of equipment were used although the vehicles had circular cross-section. That is, the problem of building a good battery (say) was solved separately from, not jointly with, the problem of building a fast vehicle. Our problem solving (decision-making) capacity is limited to only a few good solutions per manhour. To take up all problems at once is desirable but not cheap and perhaps not feasible. However, as time goes on the joint approach should be tried. Hence this economist's appeal to both statisticians and engineers.

I have just said that the limitation of the research capacity of all of us explains and possibly justifies the fact that engineers and statisticians have broken up the economics of symbol manipulation into separate sections, neglecting the essential complementarity of the several services from the point of view of the demand by the user.

However, this separation seems to be partly justified also by the economics of the services themselves; viz., by the supply side. I mean in particular the conditions for the production, and therefore for the supply, of inanimate transmission channels, such as telephones, the broadcasting apparatus, perhaps even the products of the old-fashioned printing press.

To be sure, you may not be anxious to learn about the bridal dress and be very much interested in the stock market. Yet your morning newspaper will bring you both a society page (which you will throw away) and a stock market page. Any page costs as much to print as any other page. The cost depends on the number of symbols on the page, and this corresponds to the number of bits transmitted by the printed messages. And the cost per bit turns out to be smaller if every subscriber receives both the social page and the stock market report and the sports page and the political news, regardless of his special tastes. Similarly, I am forced to subscribe to a high-fidelity telephone service although I am not interested in the music of the other person's voice. I suppose this is due to the economies of mass production. It is cheaper to produce instruments that will minimize the probability of transmission error – any error, however unimportant to me personally – than to custom-make instruments which would suit people's individual preferences. Remember that, in this country at least, with its large total demand for clothing and for food, consumers do find it advantageous to buy ready made suits and standardized groceries. To go to a Bond Street tailor or to buy fancy foods is slightly more satisfactory but so much more expensive!

The problem is familiar to operations researchers as that of optimal assortment. It is also known to social and economic statisticians, editors of census volumes, and makers of production indices. They call it "optimal aggregation." What indeed is the most economical way to break down a collection of items into groups, each to be treated as if it were homogeneous, when every detail suppressed involves some sacrifice, yet also saves cost?

Thus, it is just possible that, for the purposes of large markets (but not, I would think, for the purpose of building a particular satellite!) the isolated theory of transmission channels that minimize the probability of error – any error – is exactly what one needs. Yet, to be sure of this, we ought to have at least an approximate idea as to whether the services immediately complementing the transmission, that is, the services of coding, also exhibit advantages of mass production; and that the imperfections of available data producing and decision-making services are indeed negligible as to their economic effects.

Inanimate transmission channels do display the advantages of mass production. This makes it useful, when studying their supply conditions, to apply the pure theory of communication and to derive economically significant results from measuring information in bits. But what about other symbol-manipulating services: inquiry, coding, deciding? What can we say, in particular, about those supplied not by machines but by humans?

Before commenting on this most fascinating question, let us remind ourselves of the principles of the analysis of demand, supply, and the markets, and apply them, in particular, to the markets of symbol-manipulating services.

The demands of individual users are aggregated into the total demands for various data-providing services: total demands for weather forecasters and market prophets; for the output of research laboratories, for services of spies and detectives – given the prices of each of these services. Similarly with the total demands for various communication services – television, telephones, post office, newspapers, but also schools! – given, again, the prices of each. And so also with the demand for deciders – inventory clerks and vice-presidents for finance, and humble robots. Some of these services are substitutes for one another: for example, TV, radio and newspapers; telephone and mail. Some are mutual complements: the demand for weather data and for radio sets boosts each other. Now, to explain the "given prices" in the markets and the kind and volume of transactions that actually come about, we need to know also the supply conditions. What does it cost to produce a market survey; to print a mass-circulation paper or a professional periodical; or to run a school? And to rear and train a vice-president or to build an automatic pilot? Again, the supply conditions are interrelated, although perhaps not as closely as the demand conditions. An automatic pilot combines the services of inquiring and of deciding, and it might be more costly to produce these services separately.

At any rate, the supply of a given service or a bundle of services – at given prices! – will depend on the costs of producing various kinds and amounts of them. Under competition the price will, then, equate demand and supply.

Is this not classroom economics? Yes indeed. But it should include the more advanced parts of it which allow for oligopoly, uncertainty, and other such things, mildly called "imperfections." Particularly important are the facts of indivisibility, or more precisely, the lack of homogeneity, of standardization of many of the symbol manipulators. There exist almost unique, irreplaceable research workers, teachers, administrators; just as there exist unique choice locations for plants and harbors. The problem of unique or imperfectly standardized goods is not peculiar to the economics of inquiring, communicating, and deciding. But it has been indeed neglected in the textbooks.

Let us return to the comparison of services supplied by men and by machines. The subject has seriously worried the most creatively imaginative pessimists of science fiction – from Karel Čapek to Ray Bradbury. It has also fascinated, and has led to some serious work of, psychologists and computer scientists. The results of this work, however tentative, are of great interest to us economists.

To begin with, humans are very poor transmission channels. G. Miller, a leading psycholinguist says:

> It is an act of charity to call a man a channel at all. Compared to telephone or television channels, man is better characterized as a bottleneck. Under optimal conditions it is possible for a skilled typist or piano player to transmit 25 bits per second. ...We shall have to regard 25 bits per second as near the upper limit.

More usually, the transmission capacity of an average person in our culture is only 10 bits or less, that is, we are unable to identify a stimulus without error when the stimulus is selected from more than 2^{10}, i.e., about a thousand equiprobable alternatives (that is, when the identification logically requires at least ten yes-or-no questions). As to the so-called "short memory," an important accessory of many transmission instruments, George Miller says that "no self-respecting photographic plate would even bother to sneer at us."

But what about the other symbol-manipulating services? Take coding. The lady who is typing the almost illegible manuscript of this lecture has an uncanny gift of recognizing the intended meaning of letters and words. I think she does this mostly by looking at the context of a whole sentence, or even of the whole paper itself. This we can interpret either by saying that she has the ability of encoding almost without errors the data presented in longhand into the symbols of the typewriter alphabet; or that she decodes the longhand symbols given to her, into messages, and these into actions; viz., into pressing the keys, mostly in such a way that no error occurs. As you know, the computer industry has just begun to construct machines that may one day match the human ability to recognize simple visual patterns such as hand-written individual letters (not whole sentences!). But some people believe it will take a very long time (generations or centuries, Y. Bar-Hillel thinks) until a machine can conduct "intelligent conversation" with a man or with another machine. The key words are "heuristics" and "intuition." They are as vague as "pattern recognition," "Gestalt," and "context," and all these words are perhaps intended to have the same meaning. It is remarkable, in fact, in this very context, that you and I vaguely understand each other as to what the word "context" is intended to mean. We understand each other not letter by letter, not even word by word, but by grouping symbols into large chunks – letters (or, rather, phonemes) into words, and words into sentences, and even into larger entities, each including all sentences with the same so-called "meaning." The chunks forming the vocabulary of a human coder or decoder are of course much less numerous than the ensemble of all possible combinations of a given number of letters, say. The use of chunks diminishes therefore the flow of signals through the channel; it is more efficient, more economical. Consider the three letters C, O, W. They can be combined in six possible ways. But only one of the six combinations occurs in your vocabulary: cow. And, remarkably, it invokes not just the few properties listed by taxonomists who define a cow, but a whole image of shapes and sounds and colors and the tail waving the flies away. A most efficient, economical code – the living human language!

Most important, when you, a man, talk to a fellow human you adjust your code to the receptor, and keep readjusting it, sensitive to his feed-back responses. Is this not what characterizes a good teacher? James Boswell, young and smug, wrote in his diary:

> Health, youth, and gold sufficient I possess;
> Where one has more, five hundred sure have less.

He could as well have said: my wealth is at the upper fifth percentile. This would be economical if you would address income statisticians: you would utilize a "subroutine" that has been educated into them! But when you address other people, better recite a poem.

Even the talking to computers is better done by men, at least today. The encoding, or programming, of a difficult problem for a computer is said to be an art, not a science. People who say this probably mean precisely this: the activity of programming cannot be delegated by men to machines, at least not in serious cases and not in the present state of technology. Hence the very large proportion (one half, I understand) that the human activity of programming contributes to the value added by computing organizations.

To turn from coding to inquiring services. A biochemist (J. Lederberg) and a computer scientist (E. Feigenbaum) have told a computer how to generate the graphs of all imaginable organic compounds of a certain class, and also how to match them with certain empirically observed spectra. This was essentially a job of mathematical routine. But now comes the heuristics! The biochemist had accumulated enough experience and a flair to eliminate as unrealistic all but a few patterns from the thousands that the computer had omitted. Yet the human being, a Nobel laureate, was not able to articulate his flair, although he did learn in this direction from the co-operation with the computer. Given the abilities and the technologies, there is some optimal way of allocating the tasks between men and machines – as economists have known long ago. And we must not be too hard on the computer: its hardware is certainly much less complex than the man's genetic endowment, and the computer's short babyhood is not rich in experience.

Finally, the service of decision making or problem solving. How to allocate tasks of this nature among executives and machines? A delicate problem! It involves all echelons of a corporation, up to, but of course not including, the president, who cannot fire himself. I had better skirt this subject!

But let me remind you of the distinction I have made earlier, between decision-making and the higher-order activity of choosing who or what should provide a given service of decision making or of inquiring or of coding or transmitting. The man in charge was called the "leader" or "organizer." It is his judgment of prior probabilities and utilities, his "beliefs" and his "tastes" (or "values," in other words), that are used among the "givens" of the organizer's problem. He cannot delegate them, either to men or machines.

His problem may be, in fact, much, much larger than my chart suggests at first glance. The economic problem of organization is that of allocating numerous kinds of tasks, symbol manipulating as well as physical, to numerous transformers, arranged in a complex yet efficient network. And further complications, of a different kind, arise when a single organizer is replaced by several. Their beliefs and utilities are not the same. They engage in a nonconstant sum game. The economist's problem is then shifted from the search for optimality to the search for stability: he tries to explain, as does the biologist or anthropologist, why certain arrangements, certain allocations of tasks and incentives (rewards) have a greater chance to survive over a given period than other arrangements, and under what conditions.

The criterion of survival, viability, stability guides the social scientist who describes, and tries to explain, the existing institutions. Yet not everything that is stable is desirable. Some wicked dictatorships have been quite stable. Along with the stability criterion, the economist uses a weak collective optimality criterion, a modest common denominator on which people might agree in spite of their divergent utilities and beliefs: an arrangement of tasks and incentives is optimal in this modest sense if there is no feasible arrangement that would be better or at least not worse for all members of the organization.

What, then, if we consider the whole society as an organization? How should incentives and tasks be allocated in a way that is stable or is collectively optimal, or, if possible, both? Further, some of us cannot help but smuggle in our own values, in particular a high valuation of liberty and equity. I suppose "public policy," "public good," in our tradition, mean somehow to reconcile the criteria of stability and of collective optimality with those of liberty and equity. Though the economic theorist prefers to hide behind the technical term "welfare economics," he means not just Secretary Gardner's Department of Health, Education and Welfare, but much more, the whole public policy. Nor is our special concern only education, even if taken in the broad sense of the communication of what my chart calls "data," to the whole or some parts of the public. For research, inquiry has been also our concern here. Public policy problems in the field of symbol manipulation are crudely exemplified by questions such as, "When, if at all, should the government subsidize or protect research and teaching and the dissemination of news?"

As far as I know, welfare economics of symbol manipulation is at its beginning. Special problems, such as the theory of patents and of public versus private broadcasting and, most importantly, of the economics of education, have been studied and the names of Silberston, Coase, Gary Becker come to mind.

On the more abstract level, a basic distinction exists between the information about external facts and the information conveyed to a member of society about the doings of others. A preliminary analysis of economic policy of information about external facts has been made by my colleague Hirshleifer. If correct, his conclusions on teaching and research are quite relevant to the California battle of tuition fees, although Hirshleifer's analysis had to be based on some extreme, simplifying assumptions. To analyze the economics of information of people about other people is even harder. Game theorists have provided some building blocks. Ożga has worked on "imperfect markets through lack of knowledge" and Stigler on the information in the labor market. It is just one year ago that Leijonhufvud told this Association that Keynesian unemployment may be mostly due to lack of information. We know very little about the technology of such information; for example, about the optimal language. Indeed, many believe that the run on gold is dammed, not by verbal announcements in English or even in French, but by actually selling gold to all comers. And Radner has penetratingly pointed to the setup cost of information which makes for increasing returns to scale and makes it difficult to apply the classical theory of free markets, which reconciles optimality and stability.

All this discussion, mostly by young members of our Association, is very recent, very exciting, and, I believe, very important. The informational revolution is upon us, and the manipulation of symbols dominates our lives more and more. I do hope we shall soon understand how to harness and benefit from those trends in our culture.

Extra-terrestrial relays

Arthur C. Clarke

Although it is possible, by a suitable choice of frequencies and routes, to provide telephony circuits between any two points or regions of the earth for a large part of the time, long-distance communication is greatly hampered by the peculiarities of the ionosphere, and there are even occasions when it may be impossible. A true broadcast service, giving constant field strength at all times over the whole globe would be invaluable, not to say indispensable, in a world society.

Unsatisfactory though the telephony and telegraph position is, that of television is far worse, since ionospheric transmission cannot be employed at all. The service area of a television station, even on a very good site, is only about a hundred miles across. To cover a small country such as Great Britain would require a network of transmitters, connected by coaxial lines, waveguides or VHF relay links. A recent theoretical study has shown that such a system would require repeaters at intervals of fifty miles or less. A system of this kind could provide television coverage, at a very considerable cost, over the whole of a small country. It would be out of the question to provide a large continent with such a service, and only the main centres of population could be included in the network.

The problem is equally serious when an attempt is made to link television services in different parts of the globe. A relay chain several thousand miles long would cost millions, and transoceanic services would still be impossible. Similar considerations apply to the provision of wide-band frequency modulation and other services, such as high-speed facsimile which are by their nature restricted to the ultra-high-frequencies.

Many may consider the solution proposed in this discussion too far-fetched to be taken very seriously. Such an attitude is unreasonable, as everything envisaged here is a logical extension of developments in the last ten years – in particular the perfection of the long-range rocket of which V2 was the prototype. While this article was being written, it was announced that the Germans were considering a similar project, which they believed possible within fifty to a hundred years.

Before proceeding further, it is necessary to discuss briefly certain fundamental laws of rocket propulsion and "astronautics." A rocket which achieved a sufficiently great speed in flight outside the earth's atmosphere would never return. This "orbital" velocity is 8 km per sec. (5 miles per sec.), and a rocket which attained it would become an artificial satellite, circling the world for ever with no expenditure of power – a second moon, in fact.

The German transatlantic rocket A10 would have reached more than half this velocity.

It will be possible in a few more years to build radio controlled rockets which can be steered into such orbits beyond the limits of the atmosphere and left to broadcast scientific information back to the earth. A little later, manned rockets will be able to make similar flights with sufficient excess power to break the orbit and return to earth.

There are an infinite number of possible stable orbits, circular and elliptical, in which a rocket would remain if the initial conditions were correct. The velocity of 8 km/sec. applies only to the closest possible orbit, one just outside the atmosphere, and the period of revolution would be about 90 minutes. As the radius of the orbit increases the velocity decreases, since gravity is diminishing and less centrifugal force is needed to balance it. Fig. 1 shows this graphically. The moon, of course, is a particular case and would lie on the curves of Fig. 1 if they were produced. The proposed German space-stations would have a period of about four and a half hours.

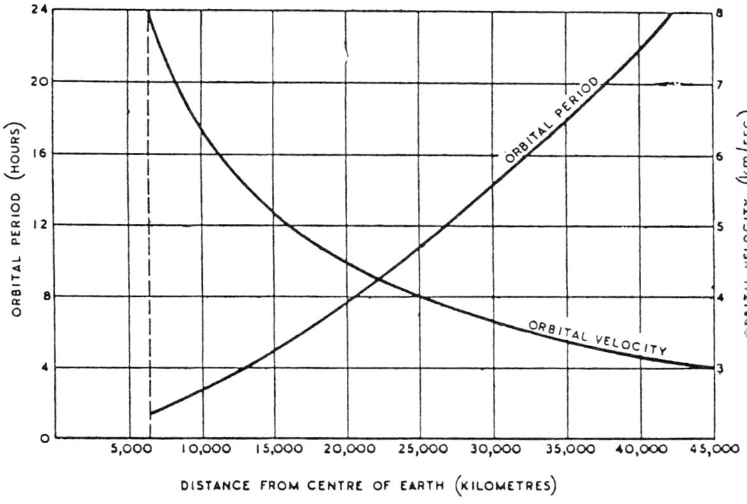

Fig. 1. Variation of orbital period and velocity with distance from the centre of the earth

It will be observed that one orbit, with a radius of 42,000 km has a period of exactly 24 hours. A body in such an orbit, if its plane coincided with that of the earth's equator, would revolve with the earth and would thus be stationary above the same spot on the planet. It would remain fixed in the sky of a whole hemisphere and unlike all other heavenly bodies would neither rise nor set. A body in a smaller orbit would revolve more quickly than the earth and so would rise in the west, as indeed happens with the inner moon of Mars.

Using material ferried up by rockets, it would be possible to construct a "space-station" in such an orbit. The station could be provided with living quarters, laboratories and everything needed for the comfort of its crew, who would be relieved and provisioned by a regular rocket service. This project might be undertaken for purely scientific reasons as it would contribute enormously to our knowledge of astronomy, physics and meteorology. A good deal of literature has already been written on the subject.

Although such an undertaking may seem fantastic, it requires for its fulfilment rockets only twice as fast as those already in the design stage. Since the gravitational stresses involved in the structure are negligible, only the very lightest materials would be necessary and the station could be as large as required.

Let us now suppose that such a station were built in this orbit. It could be provided with receiving and transmitting equipment (the problem of power will be discussed later) and could act as a repeater to relay transmissions between any two points on the hemisphere beneath, using any frequency which will penetrate the ionosphere. If directive arrays were used, the power requirements would be very small, as direct line of sight transmission would be used. There is the further important point that arrays on the earth, once set up, could remain fixed indefinitely.

Moreover, a transmission received from any point on the hemisphere could be broadcast to the whole of the visible face of the globe, and thus the requirements of all possible services would be met.

It may be argued that we have as yet no direct evidence of radio waves passing between the surface of the earth and outer space; all we can say with certainty is that the shorter wavelengths are not reflected back to the earth. Direct evidence of field strength above the earth's atmosphere could be obtained by V2 rocket technique, and it is to be hoped that someone will do something about this soon as there must be quite a surplus stock somewhere! Alternatively, given sufficient transmitting power, we might obtain the necessary evidence by exploring for echoes from the moon. In the meantime we have visual evidence that frequencies at the optical end of the spectrum pass through with little absorption except at certain frequencies at which resonance effects occur. Medium high frequencies go through the E layer twice to be reflected from the F layer and echoes have been received from meteors in or above the F layer. It seems fairly certain that frequencies from, say, 50 Mc/s to 100,000 Mc/s could be used without undue absorption in the atmosphere or the ionosphere.

A single station could only provide coverage to half the globe, and for a world service three would be required, though more could be readily utilised. Fig. 2 shows the simplest arrangement. The stations would be arranged approximately equidistantly around the earth, and the following longitudes appear to be suitable:

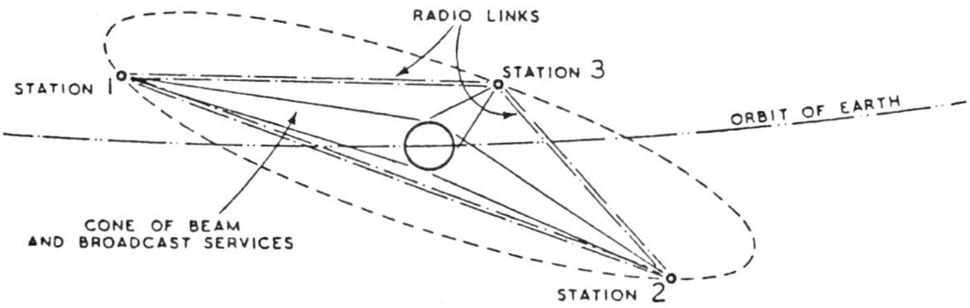

Fig. 2. Three satellite stations would ensure complete coverage of the globe

30 E – Africa and Europe
150 E – China and Oceana
90 W – The Americas

The stations in the chain would be linked by radio or optical beams, and thus any conceivable beam or broadcast service could be provided.

The technical problems involved in the design of such stations are extremely interesting, but only a few can be gone into here. Batteries of parabolic reflectors would be provided, of apertures depending on the frequencies employed. Assuming the use of 3,000 Mc/s waves, mirrors about a metre across would beam almost all the power on to the earth. Larger reflectors could be used to illuminate single countries or regions for the more restricted services, with consequent economy of power. On the higher frequencies it is not difficult to produce beams less than a degree in width, and, as mentioned before, there would be no physical limitations on the size of the mirrors. (From the space station, the disc of the earth would be a little over 17 degrees across). The same mirrors could be used for many different transmissions if precautions were taken to avoid cross modulation.

It is clear from the nature of the system that the power needed will be much less than that required for any other arrangement, since all the energy radiated can be uniformly distributed over the service area, and none is wasted. An approximate estimate of the power required for the broadcast service from a single station can be made as follows:–

The field strength in the equatorial plane of a $\lambda/2$ dipole in free space at a distance of d metres is

$$e = 6.85 \frac{\sqrt{P}}{d} \text{ volts/metre, where P is the power radiated in watts.}$$

Taking d as 42,000 km (effectively it would be less), we have $P = 37.6\ e^2$ watts. (e now in μV/metre.)

If we assume e to be 50 microvolts/metre, which is the F.C.C. standard for frequency modulation, P will be 94 kW. This is the power required for a single dipole, and not an array which would concentrate all the power on the earth. Such an array would have a gain over a simple dipole of about

80. The power required for the broadcast service would thus be about 1.2 kW.

Ridiculously small though it is, this figure is probably much too generous. Small parabolas about a foot in diameter would be used for receiving at the earth end and would give a very good signal/noise ratio. There would be very little interference, partly because of the frequency used and partly because the mirrors would be pointing towards the sky which could contain no other source of signal. A field strength of 10 microvolts/metre might well be ample, and this would require a transmitter output of only 50 watts.

When it is remembered that these figures relate to the broadcast service, the efficiency of the system will be realised. The point to-point beam transmissions might need powers of only 10 watts or so. These figures, of course, would need correction for ionospheric and atmospheric absorption, but that would be quite small over most of the band. The slight falling off in field strength due to this cause towards the edge of the service area could be readily corrected by a non-uniform radiator.

The efficiency of the system is strikingly revealed when we consider that the London Television service required about 3 kW average power for an area less than fifty miles in radius.

A second fundamental problem is the provision of electrical energy to run the large number of transmitters required for the different services. In space beyond the atmosphere, a square metre normal to the solar radiation intercepts 1.35 kW of energy.

Solar engines have already been devised for terrestrial use and are an economic proposition in tropical countries. They employ mirrors to concentrate sunlight on the boiler of a low-pressure steam engine. Although this arrangement is not very efficient it could be made much more so in space where the operating components are in a vacuum, the radiation is intense and continuous, and the low-temperature end of the cycle could be not far from absolute zero. Thermo-electric and photo-electric developments may make it possible to utilise the solar energy more directly.

Though there is no limit to the size of the mirrors that could be built, one fifty metres in radius would intercept over 10,000 kW and at least a quarter of this energy should be available for use.

The station would be in continuous sunlight except for some weeks around the equinoxes, when it would enter the earth's shadow for a few minutes every day. For this calculation, it is legitimate to consider the earth as fixed and the sun as moving round it. The station would graze the earth's shadow on the last day in February. Every day, as it made its diurnal revolution, it would cut more deeply into the shadow, undergoing its period of maximum eclipse on March 21st. On that day it would only be in darkness for 1 hour 9 minutes. From then onwards the period of eclipse would shorten, and after April 11th the station would be in continuous sunlight again until the same thing happened six months later at the autumn equinox, between September 12th and October 14th. The total period of darkness would be about two days per year, and as the longest period of eclipse would be little more than an hour there should be no difficulty in storing enough power for an uninterrupted service.

Conclusion

Briefly summarised, the advantages of the space station are as follows:

(1) It is the only way in which true world coverage can be achieved for all possible types of service.

(2) It permits unrestricted use of a band at least 100,000 Mc/s wide, and with the use of beams an almost unlimited number of channels would be available.

(3) The power requirements are extremely small since the efficiency of "illumination" will be almost 100 per cent. Moreover, the cost of the power would be very low.

(4) However great the initial expense, it would only be a fraction of that required for world networks replaced, and running costs would be incomparably less.

Appendix – Rocket design

The development of rockets sufficiently powerful to reach "orbital" and even "escape" velocity is now only a matter of years. The following figures may be of interest in this connection.

The rocket has to acquire a final velocity of 8 km/sec. Allowing 2 km/sec. for navigational corrections and air resistance loss (this is legitimate as all space-rockets will be launched from very high country) gives a total velocity needed of 10 km/sec. The fundamental equation of rocket motion is

$$V = v \, log_e R$$

where V is the final velocity of the rocket, v the exhaust velocity and R the ratio of initial mass to final mass (payload plus structure). So far v has been about 2–2.5 km/sec for liquid fuel rockets but new designs and fuels will permit of considerably higher figures. (Oxy-hydrogen fuel has a theoretical exhaust velocity of 5.2 km/sec and more powerful combinations are known.) If we assume v to be 3.3 km/sec, R will be 20 to 1. However, owing to its finite acceleration, the rocket loses velocity as a result of gravitational retardation. If its acceleration (assumed constant) is α metres/sec., then the necessary ratio R_g is increased to

$$R_g = R \quad \frac{\alpha + g}{\alpha}$$

For an automatically controlled rocket α would be about $5g$ and so the necessary R would be 37 to 1. Such ratios cannot be realised with a single rocket but can be attained by "step-rockets", while very much higher ratios (up to 1,000 to 1) can be achieved by the principle of "cellular construction".

Epilogue – Atomic power

The advent of atomic power has at one bound brought space travel half a century nearer. It seems unlikely that we will have to wait as much as twenty years before atomic-powered rockets are developed, and such rockets could reach even the remoter planets with a fantastically small fuel/mass ratio – only a few per cent. The equations developed in the appendix still hold, but v will be increased by a factor of about a thousand.

In view of these facts, it appears hardly worthwhile to expend much effort on the building of long-distance relay chains. Even the local networks which will soon be under construction may have a working life of only 20–30 years.

A mathematical theory of communication

C. E. Shannon

Introduction

The recent development of various methods of modulation such as PCM and PPM which exchange bandwidth for signal-to-noise ratio has intensified the interest in a general theory of communication. A basis for such a theory is contained in the important papers of Nyquist and Hartley on this subject. In the present paper we will extend the theory to include a number of new factors, in particular the effect of noise in the channel, and the savings possible due to the statistical structure of the original message and due to the nature of the final destination of the information.

The fundamental problem of communication is that of reproducing at one point either exactly or approximately a message selected at another point. Frequently the messages have *meaning;* that is they refer to or are correlated according to some system with certain physical or conceptual entities. These semantic aspects of communication are irrelevant to the engineering problem. The significant aspect is that the actual message is one *selected from a set* of possible messages. The system must be designed to operate for each possible selection, not just the one which will actually be chosen since this is unknown at the time of design.

If the number of messages in the set is finite then this number or any monotonic function of this number can be regarded as a measure of the information produced when one message is chosen from the set, all choices being equally likely. As was pointed out by Hartley the most natural choice is the logarithmic function. Although this definition must be generalized considerably when we consider the influence of the statistics of the message and when we have a continuous range of messages, we will in all cases use an essentially logarithmic measure.

The logarithmic measure is more convenient for various reasons:

1. It is practically more useful. Parameters of engineering importance such as time, bandwidth, number of relays, etc., tend to vary linearly with the logarithm of the number of possibilities. For example, adding one relay to a group doubles the number of possible states of the relays. It adds 1 to the base 2 logarithm of this number. Doubling the time roughly squares the number of possible messages, or doubles the logarithm, etc.

2. It is nearer to our intuitive feeling as to the proper measure. This is closely related to (1) since we intuitively measure entities by linear comparison with common standards. One feels, for example, that two punched cards should have twice the capacity of one for information

storage, and two identical channels twice the capacity of one for transmitting information.

3. It is mathematically more suitable. Many of the limiting operations are simple in terms of the logarithm but would require clumsy restatement in terms of the number of possibilities.

The choice of a logarithmic base corresponds to the choice of a unit for measuring information. If the base 2 is used the resulting units may be called binary digits, or more briefly *bits*, a word suggested by J. W. Tukey. A device with two stable positions, such as a relay or a flip-flop circuit, can store one bit of information. N such devices can store N bits, since the total number of possible states is 2^N and $\log_2 2^N = N$. If the base 10 is used the units may be called decimal digits. Since

$$\log_2 M = \log_{10} M / \log_{10} 2$$

$$= 3.32 \log_{10} M,$$

a decimal digit is about 3⅓ bits. A digit wheel on a desk computing machine has ten stable positions and therefore has a storage capacity of one decimal digit. In analytical work where integration and differentiation are involved the base e is sometimes useful. The resulting units of information will be called natural units. Change from the base a to base b merely requires multiplication by $\log_b a$.

[...]

Choice, uncertainty and entropy

[...]

Suppose we have a set of possible events whose probabilities of occurrence are $p_1, p_2, \ldots, p_n$. These probabilities are known but that is all we know concerning which event will occur. Can we find a measure of how much "choice" is involved in the selection of the event or of how uncertain we are of the outcome?

If there is such a measure, say $H(p_1, p_2, \ldots, p_n)$, it is reasonable to require of it the following properties:

1. H should be continuous in the p_i.

2. If all the p_i are equal, $p_i = \frac{1}{n}$..., then H should be a monotonic increasing function of n. With equally likely events there is more choice, or uncertainty, when there are more possible events.

3. If a choice be broken down into two successive choices, the original H should be the weighted sum of the individual values of H. The meaning of this is illustrated in Fig. 1. At the left we have three possibilities $p_1 = \frac{1}{2}$, $p_2 = \frac{1}{3}$, $p_3 = \frac{1}{6}$. On the right we first choose between two possibilities each with probability ½, and if the second occurs make another choice with probabilities ⅔, ⅓. The final results have the same probabilities as before. We require. in this special case, that

$$H(\tfrac{1}{2}, \tfrac{1}{3}, \tfrac{1}{6}) = H(\tfrac{1}{2}, \tfrac{1}{2}) + \tfrac{1}{2}H(\tfrac{2}{3}, \tfrac{1}{3})$$

The coefficient ½ is because this second choice only occurs half the time.

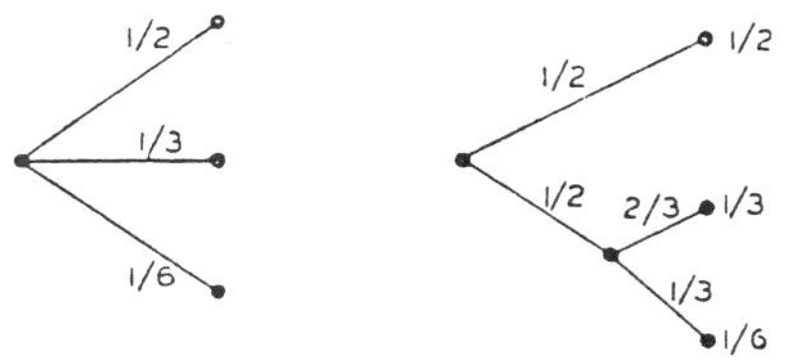

Fig. 1 – Decomposition of a choice from three possibilities.

[...] The only H satisfying the three above assumptions is of the form:

$$H = -K \sum_{i=1} p_i \log p_i$$

where K is a positive constant.

This theorem, and the assumptions required for its proof, are in no way necessary for the present theory. It is given chiefly to lend a certain plausibility to some of our later definitions. The real justification of these definitions, however, will reside in their implications.

[...]

Discussion on coding and entropy

In order to obtain the maximum power transfer from a generator to a load a transformer must in general be introduced so that the generator as seen from the load has the load resistance. The situation here is roughly analogous. The transducer which does the encoding should match the source to the channel in a statistical sense. The source as seen from the channel through the transducer should have the same statistical structure as the source which maximizes the entropy in the channel. [...] [A]lthough an exact match is not in general possible, we can approximate it as closely as desired. The ratio of the actual rate of transmission to the capacity C may be called the efficiency of the coding system. This is of course equal to the ratio of the actual entropy of the channel symbols to the maximum possible entropy.

In general, ideal or nearly ideal encoding requires a long delay in the transmitter and receiver. In the noiseless case which we have been considering, the main function of this delay is to allow reasonably good matching of probabilities to corresponding lengths of sequences. With a good code the logarithm of the reciprocal probability of a long message must be proportional to the duration of the corresponding signal, in fact

$$\frac{\log p^{-1}}{T} - C$$

must be small for all but a small fraction of the long messages.

If a source can produce only one particular message its entropy is zero, and no channel is required. For example, a computing machine set up to calculate the successive digits of π produces a definite sequence with no chance element. No channel is required to "transmit" this to another point. One could construct a second machine to compute the same sequence at the point. However, this may be impractical. In such a case we can choose to ignore some or all of the statistical knowledge we have of the source. We might consider the digits of π to be a random sequence in that we construct a system capable of sending any sequence of digits. In a similar way we may choose to use some of our statistical knowledge of English in constructing a code, but not all of it. In such a case we consider the source with the maximum entropy subject to the statistical conditions we wish to retain. The entropy of this source determines the channel capacity which is necessary and sufficient. In the π example the only information retained is that all the digits are chosen from the set 0, 1, ..., 9. In the case of English one might wish to use the statistical saving possible due to letter frequencies, but nothing else. The maximum entropy source is then the first approximation to English and its entropy determines the required channel capacity.

Examples

As a simple example of some of these results consider a source which produces a sequence of letters chosen from among A, B, C, D with probabilities ½, ¼, ⅛, ⅛, successive symbols being chosen independently. We have

$$\begin{aligned} H &= -\left(\tfrac{1}{2}\log\tfrac{1}{2} + \tfrac{1}{4}\log\tfrac{1}{4} + \tfrac{2}{8}\log\tfrac{1}{8}\right) \\ &= \tfrac{7}{4} \text{ bits per symbol.} \end{aligned}$$

Thus we can approximate a coding system to encode messages from this source into binary digits with an average of 7/4 binary digit per symbol. In this case we can actually achieve the limiting value by the following code:

A	0
B	10
C	110
D	111

The average number of binary digits used in encoding a sequence of N symbols will be

$$N\left(\tfrac{1}{2} \times 1 + \tfrac{1}{4} \times 2 + \tfrac{2}{8} \times 3\right) = \tfrac{7}{4}N$$

It is easily seen that the binary digits 0, 1 have probabilities ½, ½ so the H for the coded sequences is one bit per symbol. Since, on the average, we have $7/4$ binary symbols per original letter, the entropies on a time basis are the same. The maximum possible entropy for the original set is $\log 4 = 2$, occurring when A, B, C, D have probabilities $1/4$, $1/4$, $1/4$, $1/4$. Hence the relative entropy is $7/8$. We can translate the binary sequences into the original set of symbols on a two-to-one basis by the following table:

00	A'
01	B'
10	C'
11	D'

This double process then encodes the original message into the same symbols but with an average compression ratio $7/8$.
[...]

Equivocation and channel capacity

If the channel is noisy it is not in general possible to reconstruct the original message or the transmitted signal with *certainty* by any operation on the received signal E. There are, however, ways of transmitting the information which are optimal in combating noise. This is the problem which we now consider.

Suppose there are two possible symbols 0 and 1, and we are transmitting at a rate of 1000 symbols per second with probabilities $p_0 = p_1 = 1/2$. Thus our source is producing information at the rate of 1000 bits per second. During transmission the noise introduces errors so that, on the average, 1 in 100 is received incorrectly (a 0 as 1, or 1 as 0). What is the rate of transmission of information? Certainly less than 1000 bits per second since about 1 per cent of the received symbols are incorrect. Our first impulse might be to say the rate is 990 bits per second, merely subtracting the expected number of errors. This is not satisfactory since it fails to take into account the recipient's lack of knowledge of where the errors occur. We may carry it to an extreme case and suppose the noise so great that the received symbols are entirely independent of the transmitted symbols. The probability of receiving 1 is $1/2$ whatever was transmitted and similarly for 0. Then about half of the received symbols are correct due to chance alone, and we would be giving the system credit for transmitting 500 bits per second while actually no information is being transmitted at all. Equally "good" transmission would be obtained by dispensing with the channel entirely and flipping a coin at the receiving point.

Evidently the proper correction to apply to the amount of information transmitted is the amount of this information which is missing in the received signal, or alternatively the uncertainty when we have received a signal of what was actually sent. From our previous discussion of entropy as a measure of uncertainty it seems reasonable to use the conditional entropy of the message, knowing the received signal, as a measure of this missing information. This is indeed the proper definition, as we shall see later. Following this idea the rate of actual transmission, R, would be obtained by subtracting from the rate of production (i.e., the entropy of the source) the average rate of conditional entropy.

$$R = H(x) - H_y(x)$$

The conditional entropy $H_y(x)$ will, for convenience, be called the equivocation. It measures the average ambiguity of the received signal.

In the example considered above, if a 0 is received the *a posteriori* probability that a 0 was transmitted is .99, and that a 1 was transmitted is .01. These figures are reversed if a 1 is received. Hence

$$\begin{aligned} H_y(x) &= -\ [.99 \log .99 + 0.01 \log 0.01] \\ &= .081 \text{ bits/symbol} \end{aligned}$$

or 81 bits per second. We may say that the system is transmitting at a rate 1000 – 81 = 919 bits per second. In the extreme case where a 0 is equally likely to be received as a 0 or 1 and similarly for 1, the *a posteriori* probabilities are ½, ½ and

$$\begin{aligned} H_y(x) &= -\ [\tfrac{1}{2} \log \tfrac{1}{2} + \tfrac{1}{2} \log \tfrac{1}{2}] \\ &= 1 \text{ bit per symbol} \end{aligned}$$

or 1000 bits per second. The rate of transmission is then 0 as it should be. [...]

Channel capacity with an average power limitation

[...] [When] the noise is a white thermal noise and the transmitted signals are limited to a certain average power P, [...] the received signals have an average power $P + N$ where N is the average noise power. The maximum entropy for the received signals occurs when they also form a white noise ensemble since this is the greatest possible entropy for a power $P + N$ and can be obtained by a suitable choice of the ensemble of transmitted signals, namely if they form a white noise ensemble of power P. The entropy (per second) of the received ensemble is then

$$H(y) = W \log 2\pi e(P + N),$$

and the noise entropy is

$$H(n) = W \log 2\pi e N.$$

The channel capacity is

$$C = H(y) - H(n) = W \log \frac{P + N}{N}$$

Summarizing we have the following:

[...] The capacity of a channel of band W perturbed by white thermal noise of power N when the average transmitter power is P is given by

$$C = W \log \frac{P + N}{N}$$

This means of course that by sufficiently involved encoding systems we can transmit binary digits at the rate $W \log_2 \frac{P+N}{N}$ bits per second, with arbitrarily small frequency of errors. It is not possible to transmit at a higher rate by any encoding system without a definite positive frequency of errors.
[...]

If the noise is itself white, $N = N_1$ and the result reduces to the formula proved previously:

$$C = W \log\left(1 + \frac{P}{N}\right)$$

The convergence of computing and telecommunications

David Farber and Paul Baran

A major change in computer communications is taking place. As the cost of communications and computation declines, the number of new services that can be offered will increase and a large number of new industries can be expected to emerge. A battle between the potential suppliers is now under way.

This article considers: (i) the nature of the new services, (ii) their importance to the economy, and (iii) the arguments for regulatory control as proposed by the major competing suppliers. The basis for seeking regulation hinges primarily on the use of a communication component provided by a natural monopoly that in the past had been highly regulated. Rapid changes in the technological base make it difficult, if not impossible, to resolve the differences between the three major competing suppliers [American Telephone & Telegraph Company (AT&T), International Business Machines Corporation (IBM), and the small computer companies] within the existing communications regulatory framework. Considering computation and communications as distinctly separable functions does not provide the basis to resolve boundary disputes rapidly when regulatory issues are involved. This article examines the underlying technical changes, reflected in the conceptual changes, which now make the regulators' tasks difficult. It suggests that part of the present turmoil and uncertainty is a result of the growing use of digital processes within the communications industry itself – an inevitable evolution of the new technology. An appendix describes the evolution of data transmission technology.

We do not intend to propose solutions to the very difficult problems that this article raises. We do not know of any completely workable solutions. Our purpose is rather to sensitize the reader to the issues and problems that our country will face now and in the future.

The nature of the new services

This article was written by two authors 500 miles apart, using a computer communication system. Each writes on the same sheet of "paper". Each amplifies, modifies, and clarifies the words of the other. Misspellings are corrected, and the order of phrases is changed. Each author takes his turn tidying up the modified manuscript by pressing a few buttons and giving a few commands. This takes care of the nitty gritty in paper writing, gobbling up the blank spaces and even justifying the right margin. Discussions on content and outlook take place with the use of the same computer communication-based message systems. The method used to produce the article you are now reading is not tomorrow. It is today.

Tomorrow, computer communications systems will be the rule for remote collaboration, like the above example. As computer communications sytems become more powerful, more humane, more forgiving, and above all, cheaper, they will become ubiquitous. The obvious next question is, "Why bother to retype this manuscript manually, set type, and print this journal on paper?" Once the words in this article have been entered into the computer system, they are in fact "published" if a copy has been stored in a publicly accessible memory portion of the computer communications system. Think of the paper wasted in conventional periodical publications when only a small portion will ever be read.

The cost component of the computer communications cost-demand curve declines each year. Yet improvements of literally several orders of magnitude are possible on the basis of what is known already. Accelerated expansion of demand is a certainty. With increased use of computer communications, changes in our institutions will follow. The example given for publishing is only suggestive of the range of change. The immediate reaction of those in publishing to such change is to feel threatened. Insitutions rarely take kindly to the prospect of obsolescence of their present justification.

Importance to the economy

Our simple illustration of the potential in publishing could, and most likely will in time, spread across the entire economy. Even today we are an information-processing society. More gross national product is developed in the United States today in activities in the information sector than in the production of tangible goods.

We have chosen a dramatic application of computer communications solely for illustration. More important are the mundane applications: automated hotel reservations, credit checking, real-time financial transactions, access to insurance and medical records, general information retrieval, and real-time inventory control in business. A wide range of services such as these and beyond is coming. Automation of the office is almost a certainty. The only questions are "when", and "how", and "by whom?" Electronic transmission of much of the current first-class mail similarly will follow.

Change will be felt widely throughout the economy and, in a fundamental manner, by many institutions. Such change is generally perceived as crisis and rarely accepted with grace. Most people believe that change is something best absorbed by others. The interim period will be a period of inevitable discomfort to those organizations and professions that are affected.

The argument for regulation

History of potential suppliers in this information economy

Although computer communications promises to provide huge new markets for potential suppliers, there is uncertainty about who they will be. It is not simply a choice of AT&T, IBM, or a diversity of smaller computer companies. Rather, the question is which supplier will capture the largest, most profitable slice of the pie, and with the minimum degree of the unpleasantness of competition.

Each supplier arrives at its claim of legitimacy to supply the new markets by a different route. The telephone industry supplies communications and is now a regulated monopoly. (The most economical way of providing basic communication channels is on a common shared-resource basis – it is almost unarguably a "natural monopoly.") The IBM claim in this new industry is well-founded. It will provide a significant share of the new market's hardware. It is cash-rich and thus unique in the computer industry in having the financial resources to make the needed capital investments. And it is the winner in the proof-by-adversary test in the free economy. It has been tried and has won in the free marketplace. The third contender, the remainder of the computer industry, feels that IBM has won unfairly and seeks to handicap the giant. (There are few computer companies today that have not brought suit against IBM for alleged violations of anti-competitive stautes.) These smaller companies feel that IBM may exercise monopolistic power in a sector in part deemed not to be a natural monopoly and thus not regulated. This may all seem strange in a free economy. But the regulatory and institutional constraints are so treacherous that the outcome of the debate may hinge more on political adroitness than on market acceptance.

The disintegration of the boundary between suppliers

The basic premise that communications functions and computational functions are separable has been redered meaningless by a basic and pervasive change in technology. The separate "atoms" of communications and computers form a new "molecule" with properties different from those of their atomic components.

Consider the recent case in which AT&T wished to offer its subscribers the sophisticated model 40 communications terminals made by its wholly owned subsidiary, Teletype Corporation. A series of delays ensued as competitors argued that AT&T was entering the processing business; AT&T said "nonsense, it is just in the communications business." The Common Carrier Bureau of the Federal Communications Commission (FCC) decided against AT&T's position. After a short delay, the commissioners overturned the decision of their own bureau. Indecisiveness and on-again, off-again decisions at this boundary line are common. There is no longer an unambiguous dividing line upon which to reach decisions. The lawyers involved in resolving these arguments are assured of an employment stability lacking in most other sectors of the economy, as the endless suits at the interface between suppliers and the government seem to be careers rather than occasions.

The cause of the disintegration

The regulatory paradigm assumes that communications is communications, and computing is computing. Unlike oil and water, they now mix well – too well to treat separately, once combined. They simply are no longer separable into orthogonal components subject to different legal treatment. Separating communications processes is becoming an almost nonsense activity, like unscrambling an omelet. It is a task better left to philosophers than to lawyers on regulatory commissions.

This has not always been the case. The basic changes have crept upon us gradually. The forces melding these two previously separable domains are several. At a gross level, there is the economics of distributed processing. It may be cheaper in the future to provide independent computing power for those who need it, at their terminal, office, or home. But just having one's own computer is in general not adequate. There is the occasional or possibly frequent need for shared information where, irrespective of cost, access to remote shared data in real-time is mandatory. But, most important, the basic nature of the communications process is that digital bits provide the natural common form for all signal transmission and processing. The digital wave form becomes the universal package with which to send voice, data, facsimile, or anything else. Why this is so, and how it got that way, provides insight into the evolving technology and the increasing difficulty that will be felt by those who try to reach decisions on the premise that communications are separable in computer communications networks. What are the alternatives available for solving our dilemma?

Why aren't microcomputers a perfect substitute for large interconnected computers?

There is a school of thought that argues that computer communications is a chimera. There is no need to use the resources of AT&T or IBM. According to this viewpoint, computers are going to be so small and cheap in the future that we won't need to use communications lines to talk with big computers. Is the low-cost mini-, or micro-, or picocomputer the answer? Powerful computers have become cheap and will get cheaper. The explosion of the computer hobby market during the last year and a half has been remarkable. There are now 150 specialty stores selling low-cost computers in hobby kit form. More computer development work is being done by hobbyists today than by computer science researchers funded by public agencies. The hobbyists are not all amateurs. There are some whose vocation is also their avocation. They are intelligent, dedicated persons no longer blocked from entry into computer research by the cost of the equipment. And there are many more now at work (or play?) than the government could possibly support.

The reasons why access to communications capability will always be a necessary adjunct to the small computer are basically unchanged:

1) Whenever it is necessary to make infrequent use of a highly specialized resource, it is cheaper to share (why build a machine to make nails when you can buy the few you need at the corner hardware store?).

2) The output of one data processing task is usually the input for another process.

3) By agglomerating computer power as a result of the interconnection of a number of previously independent systems, applications that are currently not feasible in the data processing area can be handled. These include (i) systems which provide high availability, (ii) systems which can grow modularly as their environments grow (the problem of what to do with a system that supports $N + 1$ users arrives), and (iii) systems which match the physical organization of computing power with the corporate organization, thus relieving organizational strains and stresses.

Distributed processing technology and its fulfillment

All of the above applications can be satisfied by distributed processing, loosely defined as the use of a multiplicity of computational devices for a single processing task. Distributed processing implies that the units that compose the distributed computer can freely intercommunicate. If the distribution is other than local, then remote computer communications facilities are required.

Data-processing begets data-processing

Consider the following example suggesting the nature of distributed processing. A cash register in a grocery, really a counter computer, computes a customer's bill quickly. But it also generates the information needed by the store to order fresh turkeys today for delivery tomorrow. This same information can be used by a hypothetical turkey farmer to determine instantly whether there will be a shortage or a surplus of his product, and whether to hold out for a better price or sell his turkeys now. This information has economic value to his procurement decision. The key feedback signal of a free economy in the past has been the flow of money. More rapid signals can be obtained as an inexpensive by-product of the flow of information from other processes.

Economic examples are not the only use for computer networking and distributed processing. Equally, or more significant, are those that deal with human communications. Just as in the preparation of this article we were able to interchange ideas and manuscripts over the computer communications media interactively, so users of the low-cost computer systems of the future will want to intercommunicate for ideas, pleasure, and business. One has only to look at the meteoric growth of citizen's band radio to realize the new opportunities for communications between people in our society. Microcomputers with a near-zero cost making efficient use of a shared communication can open up entirely new forms of human communications.

The digital channel – the evolving choice for communications transmission

Before considering this and other issues in detail, it will be helful to review some of the characteristics of the digital channel and the reasons why it is becoming desirable even for the communication of basically analog signals. An analog signal is a voltage sent over the usual telephone line representative of (or an analog of) the pressure wave form of the human voice. Why should such a conceptually simple wave form be translated into a series of bits? The reason is that, once converted to digital format, it is possible for the wave form to tolerate a tremendous amount of distortion of its wave shape (the signal) and still be capable of simple reconstruction into a neat distortion-free series of pulses. This allows greater economy in the packing of information in a communications circuit and allows the use of lower quality (and cost) channels. Once in digital format, pulses that have been lost or converted into the wrong symbols are recoverable.

It is relatively easy to carry out many rapid calculations upon the data stream to detect and correct errors. (The price is a small cost for a few extra bits to serve as error detectors and correctors.) The ultimate in such distortion removal is cryptography, where a bit stream is purposely modified by a known-in-advance bit stream (called the key). One "subtracts" one binary wave form from the other and reconstructs the initial wave shape without distortion of the transmitted intelligence. An unauthorized observer sees only a binary stream with seemingly random chance of either being a "1" or a "0", where "1's" and "0's" are the names of the two states possible in the digital channel. Affordable secure transmission mandates digital transmission. The digital stream simplifies the task of transmission design. Here again, if the signal does not become too distorted, below 50 per cent of its initial value, it can always be reconstructed without errors. Digital transmissions unlike analog transmission, which requires critical adjustments merely to minimize distortion, is thus the ideal technique.

Switching

Communication networks require both transmission and switching. Switching is the process of connecting any network user to any other network. Switching, by its nature, is a digital process. In the 1930s, an operator plugged one circuit into the next of a tandem chain of connections through telephone switching offices. This is a digital process as each switch is either opened or closed, a neat binary process. The decision process of laying out switches and relays for telephony was among the earliest practical uses for Boolean or binary logic. The foundation of the logical design process in computer design owes much to the telephone industry. Moreover, the integrated circuit devices which make real time computer applications feasible are all offsprings of the transistor developed at the Bell Telephone laboratories. The telephone dial itself is a digital device. A series of pulses from the dial acts to set switches.

As computer technology matured to the point where its products were realizable as well as elegant and complex, the quaint electromechanical computers that formed the telephone switching network of the past were replaced by computer circuit logic. This evolution continues as digital transmission is now being combined with digital switching. The two separate processes in time are melding into a single integrated process. Bits going into one end of the network are thus beginning to be treated almost the same way as a bit stream in a computer, albeit with less processing.

What is the regulatory issue?

Who should be the future legitimate supplier of the digital system?

As the telephone technology slowly evolves toward an almost all-digital system in the far future, it will or can develop the technology needed to provide many of the new computer communications-based services. For example, AT&T has already developed devices for credit card transactions and sophisticated computer input-output terminals through its Teletype Corporation.

Equally comfortable in the domain of processing and perhaps even transmission is the highly competitive computer industry. During the last decade, the FCC encouraged competition in transmission for other than voice telephone services. New transmission companies and independent "value-added carriers" were authorized. The value-added carriers essentially lease telephone channels from the communications carriers, add processing, and use the leased lines so effectively that they are able to sell the joint product for a profit over and above the cost of the basic regulated channel.

Today, much of the in-fighting involves legal and regulatory actions seeking to block new service offerings and capabilities. It is, in part, an attempt to prevent the opponents from establishing beachheads by precedence. And, in part, it is a way of adding incertainty to the industry to discourage capital investment. Any new computer communications service offered today by a potential supplier is certain to be opposed by the other major contenders. IBM's plan for a communications satellite network for major business users, which bypasses AT&T Long Lines, faces opposition from both AT&T and computer companies fearful of IBM's assumed long-range intent. AT&T's offering of sophisticated terminals is met with countervailing opposition from the suppliers of computer terminals. The outcome of the competition between the suppliers will be decided, in part, by the staying power, legal resources, and political muscle each allocates to the fray. Little attention is paid to the "public interest". The term defies definition. Is the public interest the interest of the cross-subsidized residential telephone user? Is it the interest of a business which faces a reduced communications bill? Is public interest to be viewed primarily in the short term, irrespective of long-term damage to existing institutions in achieving immediate savings? There is no simple calculus to reach decisions. At present, there is a struggle between those who point out the advantages of competition versus those who hold that better use of limited capital results by virtue of monopoly. It is unclear which view will prevail. Until it becomes clear, resources will be wasted and future economically beneficial services to the nation will be delayed simply by the inability of industries to co-operate.

Summary

Since we do not know of any completely workable solutions to the problems we have considered, we shall settle for a few general observations:

1) The public availability of socially useful computer communications services is and has been held back by legal battles that are now under way between the potential suppliers.

2) No simple resolution of these issues in the near future seems likely in view of the past conceptual separation of computers and communications doctrines.

3) The current policy is to determine whether the nation shall or shall not have certain computer communications services, by the adversary process. In this process, often only the voices of the loudest adversary suppliers are heard.

4) Although there can be no certainty that better alternatives cannot be devised, we believe that such a possibility assumes a higher probability if the key actors come from the technical community sectors more representative of the future consumers.

5) If we are to have the new services that are possible, we need an approach that makes better use of the technologists' dreams and goals rather than have future prospects excessively bound by lawyers paid to preserve the interests of their clients, irrespective of any consequences.

6) We cannot be sanguine about this possibility, as technological statesmanship is too easily corrupted by the same forces that have placed us in this predicament. Furthermore, even if not corrupted, co-operation can too readily be regarded as simply collusion.

7) Although we do not have any clear answer, we do know that present approaches are not taking us where we want to go very rapidly and that alternative approaches should at least be considered.

Appendix: Advances in digital computer communications technology

We consider here the technical evolution of digital communications.

Modulation

The first digital transmission circuit was the electrical telegraph, predating the voice telephone by 35 years. Although local telephone lines can be and are used for simple on-off direct current (d-c) signals, the need to transmit signals beyond the limited distances where a d-c path exists led to the use of tone or alternating-current (a-c) transmission.

The device that converts the binary signal into an a-c signal for transmission is called a "modulator". The device that converts the a-c tone back into a properly shaped digital signal is called a "demodulator". The generally used arrangement, called a "modem" (MOdulator/DEModulator), combines both functions.

The first of the present-day-type modems was developed in the 1950s primarily by researchers at the Lincoln Laboratories of the Massachusetts Institute of Technology for transmitting digital signals between air defense sites over analog telephone lines. The efficacy of these devices has improved steadily since then as a result of improvements from a number of organizations. Today we are reaching perhaps 50 per cent of the theoretical transmission rate of the conventional telephone line. Modem costs have been declining steadily as the signal processing within the modem is itself converted to digital processing. This conversion is carried out effectively with the use of large-scale integration semiconductor chip circuits. The telephone plant in existence today provides a widespread access channel with low entry cost for many of today's computer applications.

Bandwidth availability

The figure of merit for a modem is the numbers of bits per second of digital information that can be transmitted over a conventional telephone channel. Data rates on the order of 300 bits per second (about 30 characters per second) simultaneously in both directions are readily achieved at low cost without electrical connection by means of acoustic coupling over any telephone line. Data rates as high as 9600 bits (one way) (about 960 characters per second) are achievable over most (but not all) voice-grade circuits with relatively expensive state-of-the-art technology.

Transmission carriers lease full-time circuits with greater bandwidths than voice circuits. Millions of bits per second can be transmitted over television bandwidth channels. Naturally, the tariffs for such arrangements are expensive, and the applications for such channels limited.

Multiplexing

Even though broad bandwidths are technically feasible, most applications can be satisfied by low data rates. A local telephone call or circuit rental has historically been inexpensive, whereas long-distance circuits have been expensive, the cost of a short-period terminal connection to a computer on the other side of the country, an expensive business, is equal to a long-distance voice call, albeit few bits may be exchanged.

The answer, of course, is combining many users' signals to share a single expensive long-distance channel. Many telegraph, and later, data signals were combined together on a single line. This process is called multiplexing. One form is called "frequency multiplexing". Here each user is assigned a different frequency, and the individual frequencies are separated at the receiving end by tuned filters. Alternatively, in time division multiplexing each user's bit streams are interleafed sequentially. Fundamental to the notion of multiplexing is the fact that each data user will not be transmitting most of the time. The fastest typist at a keyboard generates only a few tens of bits per second. For the hunt-and-peck typist, the result is only a few bits per second. The telephone channel can readily transmit 4800 bits per second, providing a potential for sharing.

Statistical multiplexing

When there is a cluster of data users who wish to send signals to a distant computer, a single expensive circuit can be used. To pack more users onto the line means taking advantage of the fact that, at any one instant of time, only a portion of the possible users would want to transmit. Moreover, to make the averaging even more effective, advantage can be taken of readily available digital storage to hold data and thus smooth out the peaks. "Time buffering" increases the efficiency of the multiplexing process. So devices were developed that combined signals from many users and elegantly packed them together for transmission over a single line.

Spatial multiplexing or packet switching

Multiplexing is highly effective if all users who want to intercommunicate with one another are in only two locations. If the potential users are spread all over the map, the same statistics can be utilized, although it becomes necessary to identify uniquely where the bits came from and where they are to go, and to add a little housekeeping data to detect damaged packets in transit. To minimize the processing equipment, the bits are formed into consistently organized packages called packets. Such packet switching appears on the ascendancy. Most widespread computer communications networks to be built in the future probably will use packet switching, at least in part.

Packet switching networks are now being built in the United States, Canada, Europe, and Japan. Interconnection between these networks is planned. Packet switching has several advantages: (i) a very robust structure can be built that readily permits the building of systems whose operation is much more reliable than its elements (communications lines and packet switching centers); (ii) it provides the highest degree of statistical averaging to make most effective use of the basic resource; (iii) the standard format package simplifies full effective interconnection between completely different computer systems and terminals; and (iv) no better alternative is in sight for most user-to-computer and most computer network applications.

Future directions for digital networks

Packet switching, although important, is by no means the end of the development for future computer communications systems. The telephone plant itself is moving on to a mostly digital structure in order to achieve a number of economic advantages. In the process it looks more like a computer communications network but with tremendously greater data-handling capacity. One might argue that the two systems are really the same. However, the minor differences between the two systems are certainly of major significance. An argument can even be raised that the conventional telephone set is nothing more than a "terminal" which generates and accepts commands and which receives and generates data to other such instruments, with the "data" being digitalized voices. A single digital voice channel in today's digital telephone systems carried 64,000 bits per second without a conventional modem. Clearly this is a tremendous capacity. There are future needs that could well use some of these high-capacity trunks of the telephone plant. There is a need in the distributed processing area for large bandwidths. In the case of intercomputer file transfers or access, bandwidths on the order of two to three megabits are appropriate whereas for the interchange of high-fidelity pictorial information 10 to 50 megabits are necessary.

Broad bandwidths are not needed everywhere. The same situation prevails in the local versus long-distance distribution of data as in the local versus toll distribution of voice communication; namely, the cost of local circuits is much less than that of toll circuits. In the case of local distribution of high-speed data, new systems such as the Pierce loop developed at Bell Laboratories, the Irvine ring designed at the University of California, Irvine, and the Ethernet produced by the Xerox Palo Alto Research Center have been proposed and constructed to attack the problem. These systems are marginally applicable to the distribution of such capabilities over large distances. The interfacing of such local data networks to national and international networks represents one of the present edges of computer communications technology.

Computers that talk and listen: man–machine communication by voice

James L. Flanagan

[...]

Computer voice response

Interest in talking machines stems from as far back as the ancient Greek and Roman civilizations. Only with the advent of electronics did speech synthesis pass from a curiosity to a potentially useful communication technique. Computers in partnership with the theory of sampled-data systems expanded the sophistication of speech synthesis, and established a new application; namely, voice readout of computer-stored data.

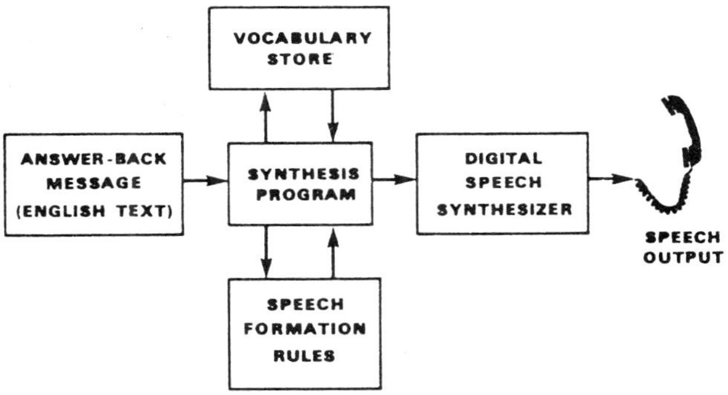

Fig. 1. Block diagram of a computer voice response system.

The ingredients of a computer voice response system are shown in Fig. 1. The machine is required to speak some message, typically specified in ordinary English text. A synthesis program must access from a stored vocabulary a description of the required sequence of words. It must put them together, head-to-tail, with the proper duration, intensity, and inflection for the prescribed context. This description of the connected utterance is then given to a synthesizer device which generates a signal for transmission over a voice circuit.

While many different approaches to speech synthesis exist, the techniques are distinguished largely by the storage capacity needed for the vocabulary and by the complexity of the control rules for generating the speech. Three different techniques serve to illustrate the range of complexity. They are, respectively, adaptive differential pulse-code modulation (ADPCM), formant synthesis, and text synthesis. Their typical bit rates for vocabulary storage are compared in Fig. 2.

Coding	*Data rate*	*Duration of speech in 10^6 bits of storage*
ADPCM	20K bits/s	1 min
Formant synthesis	500 bits/s	30 min
Text synthesis	75 bits/s	240 min

Fig. 2. Comparison of data rates for storage of speech vocabularies.

The first and simplest technique utilizes a vocabulary composed of human-spoken words whose waveforms are digitally coded. The coding method is chosen to effect an advantageous compromise among signal quality, the storage capacity required for the word vocabulary, and the simplicity of the program for assembling the output messages. This coding is exemplified by ADPCM, typically at 24 kbits/s.

Adaptive quantization applied in differential coding provides dynamic adjustment of quantizer step size according to signal characteristics, and it achieves a quality improvement equivalent to around 2 bits of quantization. As shown in Fig. 3(a), the coder and decoder have the conventional DPCM arrangement, with the exception of a logic *L*. This logic observes the bit stream produced by the coder and adjusts the quantizer step size to minimize slope overload and granular distortion. Algorithms for adaptation can involve memory for one or more past code words, and the rate of adjustment can vary between instantaneous and syllabic. Signal coding at 3 bits/sample, by an exponential logic that has one code word of memory, is illustrated in Fig. 3(b). The ADPCM is seen to provide a better estimate of the signal than does the conventional DPCM.

For message assembly, the synthesis program merely pulls the digitally coded words from disc store in the required sequence, and supplies them to an ADPCM decoder which produces the analog output signal. In this case, then, the digital synthesis device of Fig. 1 is nothing more than the simple ADPCM decoder. In this instance the only control of prosody that can be effected in assembling the message is the adjustment of time intervals between successive words, and, if necessary, their relative intensities. No control of voice pitch or merging of vocal resonances at word boundaries is possible. This technique is most appropriate, therefore, for messages requiring modest vocabulary size and having very

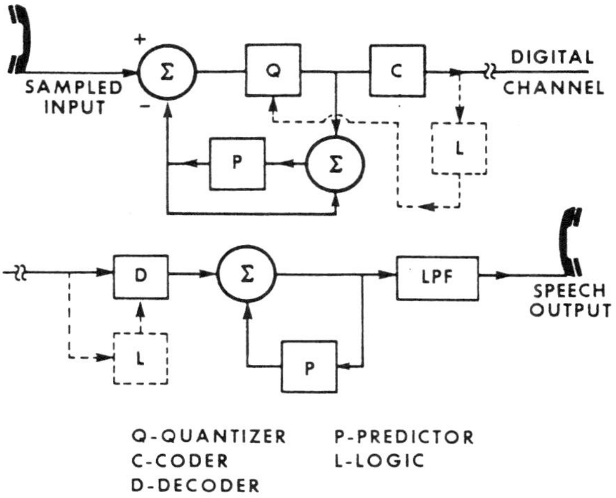

(a)

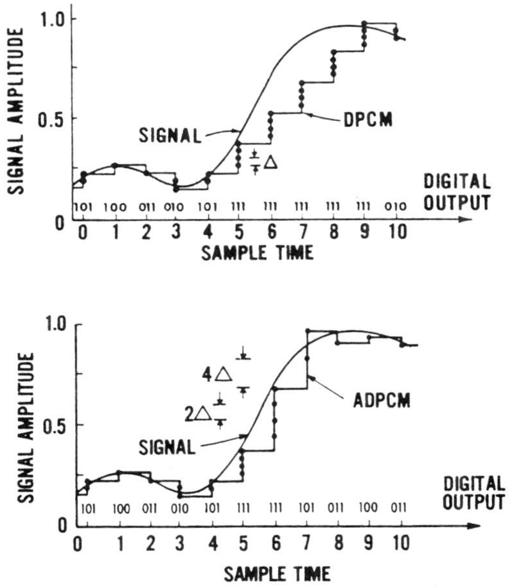

(b)

Fig. 3. Adaptive differential coding of speech. (a) Block diagram of coder (adapted from U.S. Patent 3 772 682). (b) Comparison of waveforms coded by 3-bit DPCM and ADPCM.

lax semantic constraints. Typical messages are voice readout of numerals such as telephone numbers, instructions for sequential operations such as equipment assembly or wiring, stock price quotations, and inventory reporting.

This answer-back technique is presently being used experimentally for automated directory assistance, for voice-directed wiring, and for travel information. Because the instructions for tasks such as point-to-point wiring in the assembly of electronic equipment are not highly contextual, and since the wiring list is often designed by a computer and supplied as a card deck or magnetic tape, a convenient means for converting the wiring list directly into spoken form is shown in Fig. 4. In this case, the computer voice output is recorded automatically on a cassette tape and taken to the work location. This system has been used experimentally for production-line assembly of telephone switch gear and for field installation of cabling.

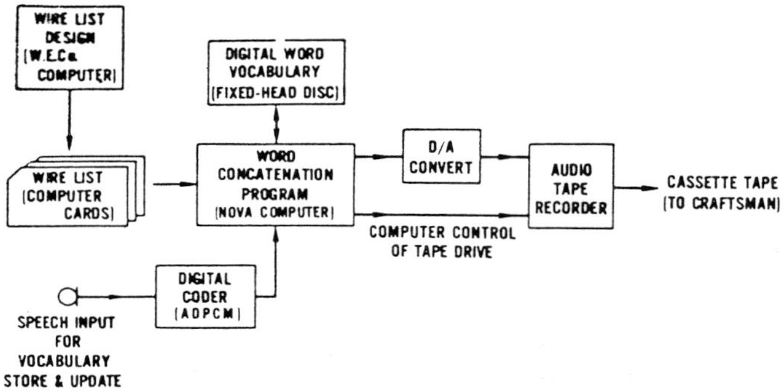

Fig. 4. Diagram of a computer system for automatically converting a wire list for telephone equipment into spoken instructions.

[...]

Field studies suggest that this voice-directed technique reduces the time required to complete the job by about 20 per cent and, perhaps more importantly, reduces the errors in wiring by about 50 per cent over the old method of visually locating and reading each instruction from a bundle of computer printout. For other types of installations, say climbing a pole to splice a cable (perhaps in a high wind), there are distinct advantages to eliminating the bundle of printout and to using a hand set to dial the computer for spoken instructions.

The second voice response technique of Fig. 2 is called formant synthesis. It utilizes a synthesis method in which a word library (again spoken initially by a human) is analyzed and parametrically stored as the time variations of vocal-tract resonances, or formants. [...]

In formant synthesis the synthesizer device is a digital filter, as shown in Fig. 5. The filter's transmission resonances and antiresonances (poles and zeroes) are controlled by the computer, and its input excitation is derived by programmed rules for voice pitch (P) and for the amplitudes of voiced sounds (A_v) and noise-like unvoiced sounds (A_n). Because of the parametric description of the speech information, this synthesis method can control and modify the prosodic characteristics (pitch, duration,

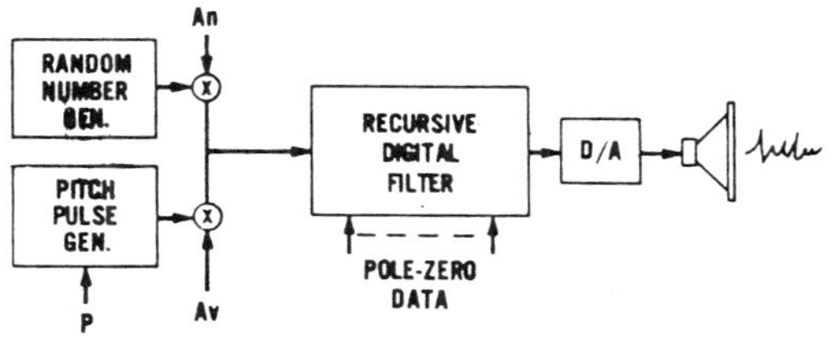

Fig. 5. Block diagram of a digital formant synthesizer.

intensity) of the synthetic speech, and it can produce smooth formant transitions at successive word boundaries. Like the ADPCM method, this technique has also been used experimentally for generating automatic-intercept messages (telephone numbers) and for producing wiring instructions by voice.

The third, and most forward-looking voice response technique mentioned in Fig. 2 is called text synthesis. It strikes at the problem of storing and voice-accessing voluminous amounts of information, or of converting any printed message into spoken form. It generates speech from an information input corresponding to a typewriter rate, about 75 bits/s. In one design, the vocabulary is literally a stored pronouncing dictionary – each entry of which (just as in an ordinary desk dictionary) has the English word, a phonetic transcription of the word, word stress marks, and some rudimentary syntax information. The synthesis program includes a syntax analyzer which examines the English message to be generated and determines the role of each word in the sentence. Using the results of parsing by the syntax analyzer, the stored rules for prosody then calculate the sound intensity, sound duration, and voice pitch appropriate for each phoneme in the particular context, as illustrated in Fig. 6.

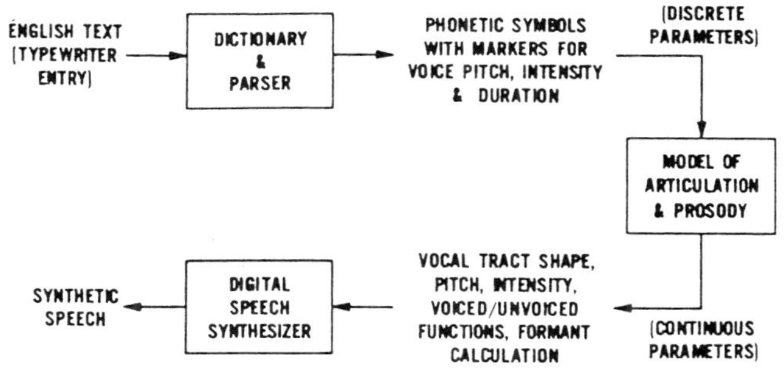

Fig. 6. System for synthesizing speech directly from printed text.

At the heart of the synthesis program is a dynamic articulatory model of the human vocal tract. The calculated controls cause the vocal-tract model to execute a sequence of "motions." These motions are described in terms of changes in the coefficients of a wave equation for sound propagation in a non-uniform tube. From the time-varying wave equation, the eigenfrequencies (or formants) of the deforming tract are computed iteratively, 100 times/s. These resonant frequencies, along with the calculated pitch and intensity information, are given to the same digital formant synthesizer shown in Fig. 5. This system therefore generates its synthetic speech completely from stored rules and without resorting to any vestige of human-produced speech. English text can be typed directly into the system, and the message synthesized on line.

Experiments in progress are examining the possibility for increasing the sophistication of the speech synthesizing device of this system. Since the system, through its model of articulation, already calculates physiologically based parameters in deriving controls for the formant synthesizer, the possibility exists for controlling a synthesis device which is more directly based upon the human speech mechanism. [...]

Speaker verification

Research over the past several years has now led to systems for authenticating the claimed identity of a caller by means of measurements on his voice signal. When the possibilities for computer voice response began to take solid form about five years ago, it also became apparent that issues would arise about accessing privileged information – for example, personal information provided by insurance clients, inventory status of manufactured products, or bank-account data to be used by designated agents. Similarly, the strong desires to reduce the amount of written material and paper exchanged in our business community gave notable impetus to the notion of a "checkless society," and to the possibilities for accomplishing banking and credit transactions by voice. Furthermore, various needs existed to streamline security procedures by means of "voice-locks" that, through on-the-spot analysis, could determine authorized access to restricted areas.

In their legitimate applications, all these situations are characterized by a co-operative individual who desires the machine to perform a useful service for him. The individual is willing to indicate his claimed identity and to have the machine judge whether the claim is correct. This is quite distinct from the problem of absolute identification of an unknown individual, or the "catch the criminal by his voice pattern" approach. Not only is the verification task more tractable technically, it is also of much more importance and utility in the business sphere.

[...] In verification the customer enters his identity claim and speaks his prearranged verification phrase. The machine makes measurements on the offered voice sample and compares the results to stored reference data it has on file for the claimed identity. It must then decide whether the claim is valid. In other words, it must generate one bit of information. Its judgment of closeness is conditioned upon its knowledge of the universe of possible users, and upon its information about the average closeness of the patterns of other users. [...] Obviously the machine can do nothing about the impostor who, having learned his closeness to a given legitimate customer, can thwart the system at will. Additionally, however, the machine's decision can be made with an error "mix" (i.e., misses versus false alarms) that is tailored to the gravity of the requested transaction.

By contrast, the task of identification involves no claimed identity from the caller, but essentially the question "Who am I?" The machine must examine its library of stored patterns and, in the absence of a more sophisticated strategy, make N comparisons and indicate the closest match. [...]

The verification problem, therefore, is judged not only more tractable, but also of more practical interest. It consequently has received more emphasis in terms of a system design. At least one system has been implemented to accomplish the functions illustrated in Fig. 7. The system uses a sentence length utterance for verification, and it has the capability for measuring the temporal patterns of voice pitch, intensity, and spectral distribution (as reflected by the speech formants, or linear-prediction coefficients). The decision threshold and error mix can be dictated by the importance of the transaction. In the clinical environment of the

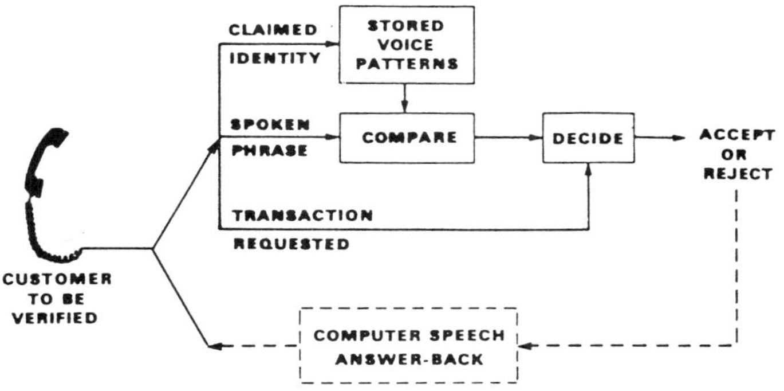

Fig. 7. Diagram of speaker verification system under test in a telephone system.

laboratory, and with low-noise high-quality speech input, the system functions with an accuracy of 99 per cent, or with a 1 per cent error, when evaluated with all customers using the same verification phrase ($N = 77$). Data for the computer when using an equal-error criterion (equal rate of false alarms and misses) is shown in Fig. 8. Of those instances (one half of the total) where the caller was a legitimate customer, it rejected him (false alarmed) 1 per cent of the time. Of those instances (also one half of the

	Reject customer	Accept impostor	Accept mimic
Computer	1	1	4
Human	4	4	22

Fig. 8. Error rates in laboratory measurements on a speaker verification system.

total) where the caller was claiming an incorrect identity, it accepted the impostor (missed) 1 per cent of the time. An obvious related issue is how good is the human ear at making the same judgment? Using exactly the same speech data as given to the computer, a parallel experiment asked human listeners to judge whether an unknown voice was the same as or different from stored reference samples. (At least one bank has used human listening for customer verification in voice banking.) The corresponding data for humans are shown also in Fig. 8. The human listeners are seen to perform less accurately than the computer.

Finally, a further related issue is how successful might skilled mimics be at violating the system? In a third series of experiments, professional voice mimics were selected from the radio and TV broadcasting industry and asked to train on the system, practising over and over, to imitate customer voices. Under these most favorable circumstances, and with all subjects using the same phrase, the skilled mimics could increase the error scores somewhat on the computer system, to 4 per cent, and by a sizable factor in the human listener experiment, to 22 per cent.

Encouraged by these tests, this verification system has now been moved, in a simplified form, into the environment of the real telephone plant. It presently operates over dialed up connections and with conventional telephone sets having carbon microphones. Using only the measures of voice pitch and intensity, it achieves a verification accuracy of 91 per cent on a customer set of 100 persons and with several thousand potential impostors. This level of performance seems acceptably high for a broad range of business transactions, such as reading out customer bank balances and making credit transactions that fall within prescribed limits. The amount of information that needs to be stored for reference data seems manageable enough by present standards of disc capacities. For example, there are about six million holders of telephone credit cards in the United States. Something on the order of 4000 bits of information appears adequate to represent a customer reference. With state-of-the-art disc stores, therefore, only about 20 disc spindles are required to store the references for all telephone credit card customers in the country. This much capacity is commonly found in most large banks or data-processing establishments. Therefore, duplication of all customer patterns in several regions over the country seems feasible and economically attractive. This duplication would make customer patterns always accessible by a relatively short-distance call.

A continuing point of research interest is the possibility of voice classification, as opposed to storage of individual reference patterns. In this instance a customer speaks or keys a code that identifies him and describes certain distinctive attributes of his voice. He then speaks a prescribed sample, and the machine makes measurements to determine whether the indicated attributes are in fact present. This approach, if proven successful, eliminates the need for any reference storage.

Another direction of research interest relates to the dependence of the verification procedure upon speaking a prescribed text. A prescribed text must, of course, be pre-arranged and remembered, but it gives added security (first through knowledge of the specific utterance and additionally, through custom-tailoring to the customer voice). It also enables the machine to measure characteristic prosody that a speaker uses in his dialect, as well as to make physical measures on text-independent features such as the characteristic spectral shape of the particular voice.

Still a further issue speaks to the question of achieving security in depth and using extemporaneous procedures to guard adequately against accepting voice samples that might have been recorded at some earlier time. All these issues represent refinements that broaden the applications of voice verification.

Speech recognition

Every field has its Holy Grail, and automatic speech recognition (ASR) is ours. Its early manifestations took the form of the speech typewriter. The advent of computers has shifted this interest more toward the possibilities for commanding machines by voice to perform services such as computation, accounting, switching, logical functions, and voice readout of requested information. As one who, along with his immediate associates, has followed this field mostly from a distance, and who only recently has come into direct contact with it, I prefer to leave comment in depth to those who have dedicated their major efforts to it, and who have pioneered it in the computer age. Nevertheless, I will, from my narrow view and narrower experience, comment about several aspects.

A formula for success in ASR seems not unlike that for speech synthesis. To wit, be prudent enough to make something immediately useful, while at the same time chewing away on the farther-out problems. (My recent admonitions about this drew mixed reactions, but no tomatoes, in Stockholm, Sweden.) Present technology and understanding clearly support highly useful and well circumscribed applications of automatic speech recognition. These applications invariably relate either to relatively small word vocabularies (such as digits) spoken by a sizable population of users, or to rather large word vocabularies (several hundred) for speakers who allow the machine to calibrate their voices. Already being demonstrated in the market place are systems for baggage and parcel sorting, for quality control and accounting on production lines, and for voice direction of machine tools.

More ambitious projects look toward speech understanding systems that will accept and understand unrestricted connected discourse from all speakers. I am not optimistic about the latter, at least not in my lifetime, nor yours. Again my preference is to achieve by pulling back from the broader issue, and taking a smaller bite – to look toward automatic recognition of connected discourse only in a rigorously limited frame – namely, for a designated speaker (on whom the machine has trained), for a highly stylized format (say, simple declarative sentences produced from a prescribed, but sizable, vocabulary), and for a highly delimited topic area. Within this context, I believe it is possible to make admirable progress over the next decade, even to the extent of getting automatically-typed letters of first-draft quality, in stylized form, about a delimited topic, and from a machine which is calibrated to a given voice. Very likely, the fundamental understanding of syntax analysis and speech constraints now being accumulated in text synthesis will advance progress toward this recognition goal.

Representative of the here and now is digit recognition with relatively high precision across speakers. A fundamental concept is the measurement of several signal features, any one of which is relatively unreliable, but when taken in combination lead to reliable recognition decisions. One such scheme is illustrated in Fig. 9, where the relatively gross measures of zero-crossing rate, signal energy, and spectral compactness and slope are used as features for digit classification. The

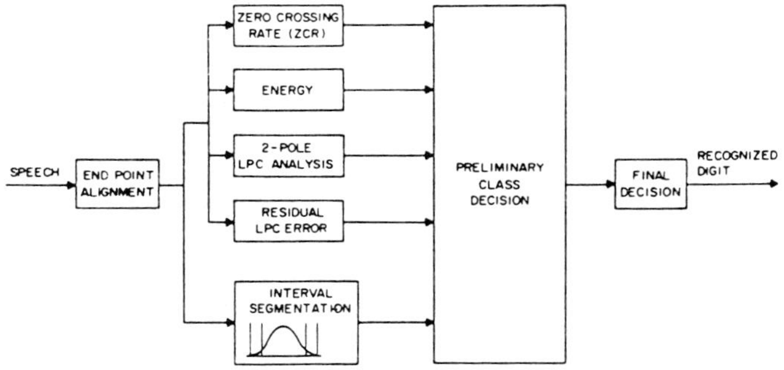

Fig. 9. System for recognition of spoken digits by feature analysis.

digit recognition scores it produces are shown in Fig. 10. [Again, alas, the male bias of this paper shows through – apologies to my female engineer and programmer colleagues for these disparate scores!]

In the direction of larger word vocabularies for designated speakers, the recognition system shown in Fig. 11 is representative. The top part of Fig. 11 shows the flow chart for the system which is based on correlation

	Number of speakers	Percent correct
Men:	25	96.8
Women:	30	92.3
Totals:	55	94.4

Fig. 10. Performance data for digit recognition.

coefficients and which utilizes a dynamic programming algorithm for time warping the measured utterance onto the reference library of the designated speaker. The lower part of Fig. 11 shows its connection in a telephone system for word recognition and reply to the customer. The system performs at a level of 96 per cent correct on a 200-word library over

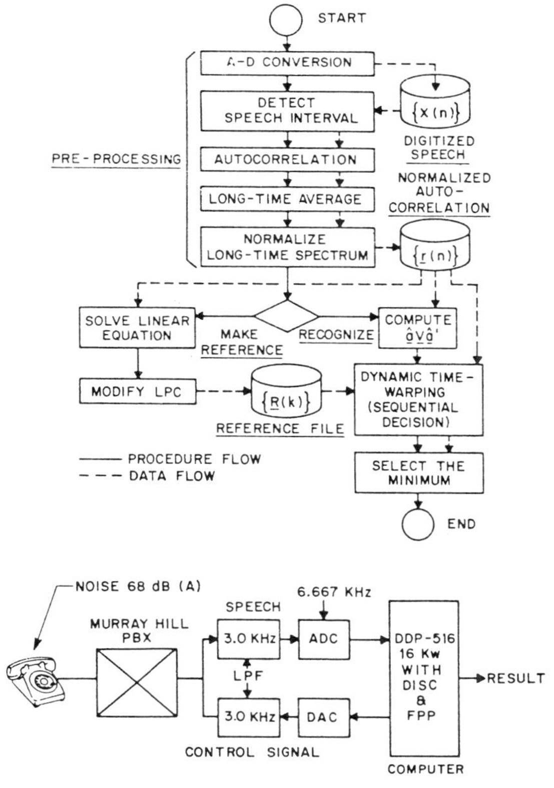

Fig. 11. Word recognition system for designated speakers.

conventional telephone connections. [...] Using conventional telephone equipment and dialed-up connections, this system is being combined with the verification and answer-back systems mentioned earlier to implement a complete "3 mode" experimental system for man-machine communication.

3-mode system for man-machine communication

Toward harnessing the speech technology we have in hand (and with the parallel intent of gaining support for frontier type activity), we have committed effort to an experimental vehicle that is an airlines information and ticketing service. This is a real-world application which at least one major airline has been considering for some time. The system being assembled in the laboratory combines 3-modes: voice answer-back based upon a digital word vocabulary coded with 24 kbits/s ADPCM, speaker verification as described earlier, and automatic recognition of a 100-word vocabulary for speakers who have trained the system. One experimental scenario is as follows:

A frequent credit customer of the airlines dials the computer. The voice response answers with:

Computer: This is ABC Airlines information and ticketing service. Please key-in your account number.
(Customer does, computer looks it up, and comes back with):
Computer: Good morning, Mr. XYZ. Please speak your verification phrase for credit charging.
(Customer does, computer churns it and, if result favorable, comes back with):
Computer: Thank you, Mr. XYZ, I confirm your charge account. Where would you like to fly to?
Customer: Washington.
Computer: From what airport would you like to fly to Washington?
Customer: Newark.
Computer: On what date would you like to fly from Newark?
Customer: (gives data; or may say no date).
Computer: At what time [on date, or of day] do you wish to depart?
Customer: (gives time).
Computer: Just a moment, please, I will give you the flights available within two hours of [time].
(Computer goes to the current airlines timetable stored in disc file, looks up flights available, reads them out, and asks):
Computer: Which flight do you wish to book?
Customer: Flight 123.
Computer: Do you wish First Class or Economy on flight 123?
Customer: First Class.
Computer: How many First Class seats do you wish?
Customer: One.

Computer: A moment, please, while I check availability of one seat.
Computer: I am pleased to confirm [reads it all back].
Your ticket will appear on your teletypewriter within the hour. Your bill will be sent at the end of the month.
Computer: Do you wish to book further?
Customer: No.
Computer: Thank you for calling ABC Airlines.
(Hangs up.)

Note two things. 1) At each stage the answerback message has a woven-in confirmation of the last recognition. If the last recognition is incorrect, the customer can push a dial button and have the previous question repeated. 2) At each stage the recognition vocabulary is contracted in size to the answer ensemble appropriate to the question. This vocabulary contraction is an important contribution towards robustness of the system.

At this writing a variant of this system is working in the laboratory over dialed-up connections and from conventional telephone sets in the local exchange. Continued tests will examine the robustness of the system, the traffic habits of the users, and the options needed. The point is that this capability exists and our present technology, properly delimited, supports it. It is a gratifying pay-off from many things learned in the interest of vocoder technology and band conservation which did not have as notable a potential.

One further point remains before concluding this commentary. A lot of the foregoing discussion has assumed an analog voice channel connecting man to computer. While, at the present time, analog-voice is far and away the most ubiquitous facility, it may not always be so. Digital voice systems and data systems will continue to the fore, and it is not too early to consider the impact of switched digital networks upon opportunities for man-machine communication.
[...]

A face to the future

What can we comfortably look toward as feasible technology within the coming five years or so?

Computer voice response is already with us, at least in rudimentary forms such as the digital word store described earlier. More sophisticated storage and readout of voluminous amounts of contextual material is also presently possible, provided one is willing to commit substantial (and expensive) computer muscle to the task, and provided one is willing to accept a great amount of automaton-quality in the signal. Knowledge about synthesis-by-rule has not yet progressed to the point of producing human-sounding speech, that is, speech which sounds like any human, much less to a quantitative description of the subtleties of individual dialect. The next several years' work should see the former better resolved, but probably not the latter.

The capabilities of automatic speaker verification now seem relatively well established. The practical precision that is possible seems applicable to a middle-spectrum of business transactions. Present performance of first-generation designs achieves the 90 per cent correct level, and does it over conventional dialed-up telephone circuits. Through more sophistication in the verification procedure, and through security "in-depth" (for example, by using sequences of custom-selected speech material keyed to temporal factors), this performance probably can be pushed up by perhaps another 5 per cent. Automatic verification therefore seems most attractive not for highest gravity transactions, nor for lowest (which do not justify any added expense), but for a gamut of transactions such as department store credit purchases, voice-lock access to restricted areas, telephone credit calls, routine checking account transactions, and accessing privileged information to authorized personnel. As microprocessor capabilities proliferate in communications equipment, verification by voice should be even further facilitated.

The prospects of speaker classification seem attractive, and research effort should be worthwhile here. This problem is a good example of where modest incremental investment might have significant and immediate return – just the type of problem we usually look to include in a research "mix."

Speech recognition at its present stage is a relatively untapped resource. The next few years should see a rapid growth in the use of word- and phrase-vocabularies for the designated-speaker type of system mentioned earlier. Most especially this includes systems which can be operated over voice-grade dialed-up connections. This disciplined and highly delimited form of speech recognition should find many real-world applications. Perhaps a good current possibility is a voice repertory dialer for mobile-radio telephone. The auto-driving telephone user would no doubt find it convenient to simply speak the name of the person he wished to talk to (selected from, perhaps, a couple of hundred names stored by the micro processor in the glove compartment of his automobile) and immediately be connected.

The speech typewriter or automated secretary, on the other hand, in its full-fledged form, i.e., accepting unrestricted connected text by any speaker, may not happen this century, if ever. A more circumscribed version of the voice typewriter – for example, a machine that produces first-draft quality for a highly specific topic area and for a speaker who has trained the machine – may be possible in this decade.

As communication system designs move more toward digital coding and transmission of speech, considerable opportunities open up for services such as variable-quality signalling, privacy or encryption, error protection in fading channels, and efficiently combining signal processing, computational functions and switching functions with the communication process – in fact, it may ultimately become difficult to distinguish these functions in a given system.

What are our limitations in seeing these technologies emerge? The established and proven ideas need only good engineering exploitation and economic stimulus. The more speculative outlooks depend for their evolution upon the originality, creativeness and innovativeness of the engineers who identify and prosecute the critical problems. As always, there are fewer people who know what to do, rather than how to do it. But this, perhaps, is as it must be and should be. Our horizons are limited only by our abilities to conceive bold and innovative techniques for communication.

The scientific journal of the future

John Senders

Part of the obligation of any science is to make public the results of scientific investigation. The scientific component of printed matter has grown at an extraordinary rate even when compared with the growth of information and information flow in general in the last fifty years. However, it has been clear to many observers that scientific journals at some time are likely to become extinct merely because of the growth of the amount of information being distributed by journals and the consequent difficulty of retrieving that information from the library. Increase in prices and depletion of resources are other reasons why the journals cannot last indefinitely. For example, the costs of raw materials for publishing are increasing faster than costs in other areas of industry. The costs and delays associated with the postal system in the presence of a rapidly increasing population have themselves changed (for the worse) more rapidly than many other aspects of society influenced by population size. In general, the situation will get worse, not better. Under such circumstances, the ideas of sociologists and their research and analyses are disseminated rapidly only to a small fraction of the potential readership. Only well-known authors are to be sent a preprint of the document. However, the document is usually uncitable in later work until it has been published, so that delays which may be as long as one year are built in to the successive generations of thought and research which stem from some fundamental idea.

As the printed journal distributed by the postal system becomes obsolete, a number of alternatives present themselves. One of these assumes that the journal will exist in its present form up to the point where it is printed and bound. It will then be made available in some micro form (micro-film or micro-fiche) which will make significant savings possible, both in production and postage costs. Most of the delays, however, are still present. Further, the information contained in the scientific article remains subject to the skill of the indexer in selecting key words which adequately describe the informational content, and to the skill of the potential searcher who may wish to obtain that information in translating his desire into a set of key words which will, in fact, retrieve, not the information, but the document which contains it.

Two major difficulties of the present system stem from the problems of distribution of current issues and the retrieval of past information. In addition, there are a number of conflicting goals. On the one hand, people who do research wish to communicate quickly the results of their activities to other persons. On the other hand, a system of quality control has been developed which, by a process involving editors, associate editors, referees, anonymous comments, confidential critiques, etc., prevents below-standard research, poorly written reports, or improperly analyzed data from finding a way into the public domain. The system also delays the information on its way from "source to sink" So that "users, experiencing long delays, still complain that many papers are already obsolete before they appear," and "these lags are attributable to causes: increased demands on editors to process more manuscripts; the slowness with which manuscripts, galleys, and page proofs flow through the editorial review and postal systems; the time required for technical editing and page layout; and the long, often unpredictable, time consumed by printing, binding, and mailing" (Van Cott, 1970).

From many points of view, the saving of time and possible elimination of duplication aside, cost considerations alone suggest that some alternative method is inevitable. There are, of course, many alternatives, but I will consider in detail only one form of electronic distribution of readable material.

This electronic alternative assumes that the present form of the journal will disappear sometime and be replaced by completely electronic storage and retrieval of the alpha-numeric-graphic content of scientific articles. Publication, for example, might consist of electronic transmission to all subscribers via "teletype" (not really, but the word has the right connotation). One objection to such proposals when they have been advanced in the past has been that such transmission would be slow and prohibitively expensive. However, with the development of digital data networks, the cost is trivial for a user who has a suitable terminal in his institution. We can easily tolerate transmission cost, especially when all of the cost previously incurred in the gathering up of trees, and conversion of them into paper, the printing and binding of the paper, and the mailing of the printed journals to various places are eliminated altogether. The question of past information with this form of publication system is dealt with later on.

Now let us examine the second goal: the preservation of quality. Consider the scientific communication process as it presently exists. Someone does a piece of research, gathers data, performs analysis, and then writes a paper, submits it to an editor and it is published. That is the simple way of looking at it. In reality things are somewhat different. The author gathers and analyzes his data; he writes a paper; he reads it and he rewrites it; he edits it and revises it again. Then he submits it to a colleague. The colleague looks at it with a critical eye, and returns it with comments. The author then rewrites or revises it and, finally, sends it off to a journal. The article is mailed to the editor who puts it at the bottom of the pile of papers awaiting his attention. Eventually, when he gets around to reading the title and the abstract (and possibly the whole text), he chooses an associate editor, or one of his referees, and mails it off to him. The referee may or may not agree to deal with this particular article. Let us assume that he does. He then has to find time to read and review it. These are difficult and time-consuming processes. The referee sends it back to the editor; the editor again puts it at the bottom of his "push-up list"; eventually he gets to it, looks at it, and sends it back to the author. If the author complies with all the suggestions made by the referees, the editor publishes the paper. To publish it, he sends it to the printer; the printer sets it in type, and one can imagine the rest. Eventually, after the galleys have been sent back, etc., the paper finds its way into the rather limited number of pages available to that journal for that year, and copies of that journal are distributed. The whole process may take one year, and sometimes up to two years.

The advantage of this system is that the articles that are so carefully processed and finally published in the highly respected journals are reliable work. The electronic alternative proposed is a system which can allow both the rapid transmission of large quantities of information from one place to another and still maintain a close control over the quality of the information which is transmitted.

In contrast to traditional journals, the electronic journal appears to have a relatively constant to slowly declining cost over the next twenty-five years. As Hecht (1976) writes: "Although the initial cost for an Electronic Journal is greater than that for paper journals, the projected cost pairs cross over during the next twenty-five years. The exact date of the cross-over depends on the number of printed pages of material. A single journal of under 1000 pages (in 1975) with 3,700 subscribers has an estimated cross-over point in the late 1990s. The entire English language scientific journal output distributed world-wide reached a cross-over at the beginning of 1976." Intermediate numbers of assumed annual pages and subscribers yield intermediate cost cross-over points.

Let us consider the following scenario: The author goes through the same internal process in his institution. He writes, and rewrites, and edits, and submits for local criticism. He finally approves the copy; that is to say, he gets it to his secretary, or to the departmental secretary. He also selects an editor. The final typing is done on a machine that also makes a punched paper tape, for example. Having selected the editor, he puts the tape in the tape reader, types in the name of the editor, and through a variety of several processes (which really are not very difficult or magical at all) he sends the content to that editor. The title, abstract, and other useful information appear that day, or three minutes after midnight, or whenever, on the editor's display. In the morning when the editor comes in, he sees all the information there. After looking at the title and the abstract which have appeared, and at the institution and the author's name, he then may choose an associate editor, or two or three, and, using very simple typed commands, send the article to the associate editors. (If he desires text, he "calls" for it and it will be displayed for him.)

The title and the abstract appear on the associate editor's display. The associate editor either accepts or refuses to referee it. Let's assume he accepts. He then goes through the same process. He calls for the text, the figures, and everything else that goes into the article, and follows exactly the same procedures as he ordinarily does, but meanwhile we are already three or four weeks (perhaps even three or four months) ahead of the usual process. The associate editor makes his comments. (By the way, each line is numbered – a trivial exercise for the computer – so that any line can be referred to by number in the commentary of the associate editor.) Having completed his commentary, he transmits this back to the editor, merely by typing in "Editor," and it appears on the editor's machine as a commentary, which the editor can then transmit (and he doesn't have to read it unless he wants to, though he more than likely will) back to the author. This, of course, can be repeated many times; but if the author complies, that is to say, rewrites in accordance with the considered judgment of the external referees, as perhaps filtered by the editor, he resubmits this material by merely transmitting it again. (The machine, by the way, will do all the interpolation, and will reorganize and renumber the lines, a simple action for the computer.) If the editor is satisfied, he types "Publish" and it is thereby published.

Now what does it mean to say "It is thereby published"? It means that the entire document of the scientific work has been entered into a magnetic store, numbered, with an author and a title, and an abstract or summary. The citation would also appear upon your display, since you subscribe to articles of that kind, i.e., the *keywords* are among those listed by you on your interest profile. If you want a full text, you get it, or you may call for any part of the complete information. For example, you may wish to call for the bibliography. (You want to know if you are cited; and if you are not, you immediately conclude that it is a mediocre article: its author wasn't aware that you had done the same work 16 years before and have been waiting for someone to cite you ever since.) You may want the "table of contents," or the summary, or just the conclusions – if your interest is in the application of the results. You may also ask for the raw data since you may have an alternative theoretical explanation or approach. Alternatively you may want the author's bibliography, since if this is an interesting article the author may have written other interesting papers. As can be seen, browsing becomes much more productive than was previously possible without a staff of library researchers at command.

The electronic search and retrieval systems based upon computer readable bibliographic citation – and in some cases, an abstract content – already exist in a number of fields. Data bases containing information on physical and social sciences are accessible from any point in the world by persons with suitable terminal equipment. For modest cost, these data bases can be searched in various ways to yield useful, bibliographic information for the person doing research or investigating a particular problem. The use of computers for editorial processing is already well under development by the National Science Foundation,

Electronic Editorial Processing centers for scientific journals are being tested experimentally at the present time. The outcome is a computer-ready tape which, in the usual course of events, would go directly into a computer-controlled typesetter in order to set hot type, or phototype for use in the traditional printing process. Obviously enough, with a computer readable output stemming from the Electronic Editorial Processing center and with the availability of storage, search, and retrieval systems of the sort characterized by Psychological Abstracts et al., the intervening stage is a minor, albeit somewhat expensive, implementation problem. Rather than its being emitted to control a typesetting machine, one could imagine the "document," once accepted for publication, being stored in a capacious electronic memory. This memory would then be accessible from any point by "subscribers" to the particular serial publication. The "journal" would once again exist in its etymological sense; i.e., publication would be carried on every day. It would, in fact, be a continuous process, and the present form of separated subsections of the scientific literature now called journals would no longer be necessary. The search and retrieval mechanisms could be done as they have been in the past, by key words, or, with the advent of cheaper, more powerful systems, by direct consideration of the entire content of the scientific article.

Additional benefits can come from such a system. It was demonstrated in 1962 that it is possible to store large quantities of alpha-numeric and graphical material in a digital computer; to recall these in a whole text search mode; to examine the graphical and quantitative content presented in the text; to manipulate the presented material according to different hypotheses about them; in short, to engage interactively with the linguistic, the quantitative, and the mathematical material presented in a document.

Many people have expressed fears that electronic offices will generate a generation of scientists who will sit at desks connected to remote machines and engage in intercourse with these machines to the exclusion of human interaction. Vannevar Bush had some great visions more than 30 years ago. Much of what he described can now be done, perhaps all of it and more. It is instructive to examine both the original and the "revisited" versions of Memex and to consider the advantages of the systems of that degree of complexity or greater (as is now possible). Most of us spend a large amount of time reading, and almost as much time bemoaning the fact that there is so much to read that we cannot possibly read it all. In addition, we are often forced to read what we do not want to read. The advantage, and it may be the overwhelming and forceful advantage, of the electronic alternative is that it will be possible to identify and locate the particular nuggets which in the past required so much mining. The electronic office is not a fearsome thing; a brief exposure to it is very convincing.

A system like the one described, or possibly some even more elaborate system, is almost inevitable given the joint pressures of volume of published scientific material and the cost of traditional ways of doing things. It would be wise for all established scientific investigators to act now to influence and support these new systems and to shape them according to what they believe to be best for the journals of the future.

Can a machine think?

A.M. Turing

The imitation game

I propose to consider the question, 'Can machines think?' This should begin with definitions of the meaning of the terms 'machine' and 'think'. The definitions might be framed so as to reflect so far as possible the normal use of the words, but this attitude is dangerous. If the meaning of the words 'machine' and 'think' are to be found by examining how they are commonly used it is difficult to escape the conclusion that the meaning and the answer to the question, 'Can machines think?' is to be sought in a statistical survey such as a Gallup poll. But this is absurd. Instead of attempting such a definition I shall replace the question by another, which is closely related to it and is expressed in relatively unambiguous words.

The new form of the problem can be described in terms of a game which we call the 'imitation game.' It is played with three people, a man (A), a woman (B), and an interrogator (C) who may be of either sex. The interrogator stays in a room apart from the other two. The object of the game for the interrogator is to determine which of the other two is the man and which is the woman. He knows them by labels X and Y, and at the end of the game he says either 'X is A and Y is B' or 'X is B and Y is A.' The interrogator is allowed to put questions to A and B thus:

C: Will X please tell me the length of his or her hair?

Now suppose X is actually A, then A must answer. It is A's object in the game to try and cause C to make the wrong identification. His answer might therefore be:

'My hair is shingled, and the longest strands are about nine inches long.'

In order that tones of voice may not help the interrogator the answers should be written or better still, typewritten. The ideal arrangement is to have a teleprinter communicating between the two rooms. Alternatively the question and answers can be repeated by an intermediary. The object of the game for the third player (B) is to help the interrogator. The best strategy for her is probably to give truthful answers. She can add such things as 'I am the woman, don't listen to him!' to her answers, but it will avail nothing as the man can make similar remarks.

We now ask the question, 'What will happen when a machine takes the part of A in this game?' Will the interrogator decide wrongly as often when the game is played like this as he does when the game is played between a man and a woman? These questions replace our original, 'Can machines think?'

Critique of the new problem

As well as asking, 'What is the answer to this new form of the question', one may ask, 'Is this new question a worthy one to investigate?' This latter question we investigate without further ado, thereby cutting short an infinite regress.

The new problem has the advantage of drawing a fairly sharp line between the physical and the intellectual capacities of a man. No engineer or chemist claims to be able to produce a material which is indistinguishable from the human skin. It is possible that at some time this might be done, but even supposing this invention available we should feel there was little point in trying to make a 'thinking machine' more human by dressing it up in such artificial flesh. The form in which we have set the problem reflects this fact in the condition which prevents the interrogator from seeing or touching the other competitors, or hearing their voices. Some other advantages of the proposed criterion may be shown up by specimen questions and answers. Thus:

Q: Please write me a sonnet on the subject of the Forth Bridge.
A: Count me out on this one. I never could write poetry.
Q: Add 34957 to 70764.
A: (Pause about 30 seconds and then give as answer) 105621.
Q: Do you play chess?
A: Yes.
Q: I have K at my K1, and no other pieces. You have only K at K6 and R at R1. It is your move. What do you play?
A: (After a pause of 15 seconds) R–R8 mate.

The question and answer method seems to be suitable for introducing almost any one of the fields of human endeavour that we wish to include. We do not wish to penalise the machine for its inability to shine in beauty competitions, nor to penalise a man for losing in a race against an aeroplane. The conditions of our game make these disabilities irrelevant. The 'witnesses' can brag, if they consider it advisable, as much as they please about their charms, strength or heroism, but the interrogator cannot demand practical demonstrations.

The game may perhaps be criticised on the ground that the odds are weighted too heavily against the machine. If the man were to try and pretend to be the machine he would clearly make a very poor showing. He would be given away at once by slowness and inaccuracy in arithmetic. May not machines carry out something which ought to be described as thinking but which is very different from what a man does? This objection is a very strong one, but at least we can say that if, nevertheless, a machine can be constructed to play the imitation game satisfactorily, we need not be troubled by this objection.

It might be urged that when playing the 'imitation game' the best strategy for the machine may possibly be something other than imitation of the behaviour of a man. This may be, but I think it is unlikely that there is any great effect of this kind. In any case there is no intention to investigate here the theory of the game, and it will be assumed that the best strategy is to try to provide answers that would naturally be given by a man.

The machines concerned in the game

The question which we put above will not be quite definite until we have specified what we mean by the word 'machine'. It is natural that we should wish to permit every kind of engineering technique to be used in our machines. We also wish to allow the possibility that an engineer or team of engineers may construct a machine which works, but whose manner of operation cannot be satisfactorily described by its constructors because they have applied a method which is largely experimental. Finally, we wish to exclude from the machines men born in the usual manner. It is difficult to frame the definitions so as to satisfy these three conditions. One might for instance insist that the team of engineers should be all of one sex, but this would not really be satisfactory, for it is probably possible to rear a complete individual from a single cell of the skin (say) of a man. To do so would be a feat of biological technique deserving of the very highest praise, but we would not be inclined to regard it as a case of 'constructing a thinking machine.' This prompts us to abandon the requirement that every kind of technique should be permitted. We are the more ready to do so in view of the fact that the present interest in 'thinking machines' has been aroused by a particular kind of machine, usually called an 'electronic computer' or 'digital computer.' Following this suggestion we only permit digital computers to take part in our game.

This restriction appears at first sight to be a very drastic one. I shall attempt to show that it is not so in reality. To do this necessitates a short account of the nature and properties of these computers.

It may also be said that this identification of machines with digital computers, like our criterion for 'thinking,' will only be unsatisfactory if (contrary to my belief), it turns out that digital computers are unable to give a good showing in the game.

There are already a number of digital computers in working order, and it may be asked, 'Why not try the experiment straight away? It would be easy to satisfy the conditions of the game. A number of interrogators could be used, and statistics compiled to show how often the right identification was given.' The short answer is that we are not asking whether all digital computers would do well in the game nor whether the computers at present available would do well, but whether there are imaginable computers which would do well. But this is only the short answer. We shall see this question in a different light later.

[...]

Contrary views on the main question

[...]

It will simplify matters for the reader if I explain first my own beliefs in the matter. Consider first the more accurate form of the question. I believe that in about fifty years' time it will be possible to program computers, with a storage capacity of about 10^9, to make them play the imitation game so well that an average interrogator will not have more than 70 per cent chance of making the right identification after five minutes of questioning. The original question, 'Can machines think?' I believe to be too meaningless to deserve discussion. Nevertheless I believe that at the end of the century the use of words and general educated opinion will have altered so much that one will be able to speak of machines thinking without expecting to be contradicted. I believe further that no useful purpose is served by concealing these beliefs. The popular view that scientists proceed inexorably from well established fact to well-established fact, never being influenced by any unproved conjecture, is quite mistaken. Provided it is made clear which are proved facts and which are conjectures, no harm can result. Conjectures are of great importance since they suggest useful lines of research.

I now proceed to consider opinions opposed to my own.

(1) The theological objection

Thinking is a function of man's immortal soul. God has given an immortal soul to every man and woman, but not to any other animal or to machines. Hence no animal or machine can think.

I am unable to accept any part of this, but will attempt to reply in theological terms. I should find the argument more convincing if animals were classed with men, for there is a greater difference, to my mind, between the typical animate and the inanimate than there is between man and the other animals. The arbitrary character of the orthodox view becomes clearer if we consider how it might appear to a member of some other religious community. How do Christians regard the Moslem view that women have no souls? But let us leave this point aside and return to the main argument. It appears to me that the argument quoted above implies a serious restriction of the omnipotence of the Almighty. It is admitted that there are certain things that He cannot do such as making one equal to two, but should we not believe that He has freedom to confer a soul on an elephant if He sees fit? We might expect that He would only exercise this power in conjunction with a mutation which provided the elephant with an appropriately improved brain to minister to the needs of this soul. An argument of exactly similar form may be made for the case of machines. It may seem different because it is more difficult to 'swallow'. But this really only means that we think it would be less likely that He would consider the circumstances suitable for conferring a soul. The circumstances in question are discussed in the rest of this paper. In attempting to construct such machines we should not be irreverently usurping His power of creating souls, any more than we are in the procreation of children: rather we are, in either case, instruments of His will providing mansions for the souls that He creates.

However, this is mere speculation. I am not very impressed with theological arguments whatever they may be used to support. Such arguments have often been found unsatisfactory in the past. In the time of Galileo it was argued that the texts, 'And the sun stood still ... and hasted not to go down about a whole day' (Joshua x. 13) and 'He laid the foundations of the earth, that it should not move at any time' (Psalm cv. 5) were an adequate refutation of the Copernican theory. With our present knowledge such an argument appears futile. When that knowledge was not available it made a quite different impression.

(2) The 'heads in the sand' objection

'The consequences of machines thinking would be too dreadful. Let us hope and believe that they cannot do so.'

This argument is seldom expressed quite so openly as in the form above. But it affects most of us who think about it at all. We like to believe that Man is in some subtle way superior to the rest of creation. It is best if he can be shown to be *necessarily* superior, for then there is no danger of him losing his commanding position. The popularity of the theological argument is clearly connected with this feeling. It is likely to be quite strong in intellectual people, since they value the power of thinking more highly than others, and are more inclined to base their belief in the superiority of Man on this power.

I do not think that this argument is sufficiently substantial to require refutation. Consolation would be more appropriate: perhaps this should be sought in the transmigration of souls.

(3) The mathematical objection

There are a number of results of mathematical logic which can be used to show that there are limitations to the powers of discrete-state machines. The best known of these results is known as Gödel's theorem, and shows that in any sufficiently powerful logical system statements can be formulated which can neither be proved nor disproved within the system, unless possibly the system itself is inconsistent. There are other, in some respects similar, results due to Church, Kleene, Rosser, and Turing. The latter result is the most convenient to consider, since it refers directly to machines, whereas the others can only be used in a comparatively indirect argument: for instance if Gödel's theorem is to be used we need in addition to have some means of describing logical systems in terms of machines, and machines in terms of logical systems. The result in question refers to a type of machine which is essentially a digital computer with an infinite capacity. It states that there are certain things that such a machine cannot do. If it is rigged up to give answers to questions as in the imitation game, there will be some questions to which it will either give a wrong answer, or fail to give an answer at all however much time is allowed for a reply. There may, of course, be many such questions, and questions which cannot be answered by one machine may be satisfactorily answered by another. We are of course supposing for the present that the questions are of the kind to which an answer 'Yes' or 'No' is appropiate, rather than questions such as 'What do you think of Picasso?' The questions that we know the machines must fail on are of this type, 'Consider the machine specified as follows ... Will this machine ever answer "Yes" to any question?' The dots are to be replaced by a description of some machine in a standard form [...]. When the machine described bears a certain comparatively simple relation to the machine which is under interrogation, it can be shown that the answer is either wrong or not forthcoming. This is the mathematical result: it is argued that it proves a disability of machines to which the human intellect is not subject.

The short answer to this argument is that although it is established that there are limitations to the powers of any particular machine, it has only been stated, without any sort of proof, that no such limitations apply to the human intellect. But I do not think this view can be dismissed quite so lightly. Whenever one of these machines is asked the appropriate critical question, and gives a definite answer, we know that this answer must be wrong, and this gives us a certain feeling of superiority. Is this feeling illusory? It is no doubt quite genuine, but I do not think too much importance should be attached to it. We too often give wrong answers to questions ourselves to be justified in being very pleased at such evidence of fallibility on the part of the machines. Further, our superiority can only be felt on such an occasion in relation to the one machine over which we have scored our petty triumph. There would be no question of triumphing simultaneously over *all* machines. In short, then, there might be men cleverer than any given machine, but then again there might be other machines cleverer again, and so on.

Those who hold to the mathematical argument would, I think, mostly be willing to accept the imitation game as a basis for discussion. Those who believe in the two previous objections would probably not be interested in any criteria.

(4) The argument from consciousness

This argument is very well expressed in Professor Jefferson's Lister Oration for 1949, from which I quote. 'Not until a machine can write a sonnet or compose a concerto because of thoughts and emotions felt, and not by the chance fall of symbols, could we agree that machine equals brain – that is, not only write it but know that it had written it. No mechanism could feel (and not merely artificially signal, an easy contrivance) pleasure at its successes, grief when its valves fuse, be warmed by flattery, be made miserable by its mistakes, be charmed by sex, be angry or depressed when it cannot get what it wants.'

This argument appears to be a denial of the validity of our test. According to the most extreme form of this view the only way by which one could be sure that a machine thinks is to *be* the machine and to feel oneself thinking. One could then describe these feelings to the world, but of course no one would be justified in taking any notice. Likewise according to this view the only way to know that a *man* thinks is to be that particular man. It is in fact the solipsist point of view. It may be the most logical view to hold but it makes communication of ideas difficult. A is liable to believe 'A thinks but B does not' whilst B believes 'B thinks but A does not'. Instead of arguing continually over this point it is usual to have the polite convention that everyone thinks.

I am sure that Professor Jefferson does not wish to adopt the extreme and solipsist point of view. Probably he would be quite willing to accept the imitation game as a test. The game (with the player B omitted) is frequently used in practice under the name of *viva voce* to discover whether someone really understands something or has 'learnt it parrot fashion'. Let us listen in to a part of such a *viva voce*:

> Interrogator: In the first line of your sonnet which reads 'Shall I compare thee to a summer's day', would not 'a spring day' do as well or better?
> Witness: It wouldn't scan.
> Interrogator: How about 'a winter's day'. That would scan all right.
> Witness: Yes, but nobody wants to be compared to a winter's day.
> Interrogator: Would you say Mr. Pickwick reminded you of Christmas?
> Witness: In a way.
> Interrogator: Yet Christmas is a winter's day, and I do not think Mr. Pickwick would mind the comparison.
> Witness: I don't think you're serious. By a winter's day one means a typical winter's day, rather than a special one like Christmas.

And so on. What would Professor Jefferson say if the sonnet-writing machine was able to answer like this in the *viva voce*? I do not know whether he would regard the machine as 'merely artificially signalling' these answers, but if the answers were as satisfactory and sustained as in the above passage I do not think he would describe it as 'an easy contrivance'. This phrase is, I think, intended to cover such devices as the inclusion in the machine of a record of someone reading a sonnet, with appropriate switching to turn it on from time to time.

In short then, I think that most of those who support the argument from consciousness could be persuaded to abandon it rather than be forced into the solipsist position. They will then probably be willing to accept our test.

I do not wish to give the impression that I think there is no mystery about consciousness. There is, for instance, something of a paradox connected with any attempt to localise it. But I do not think these mysteries necessarily need to be solved before we can answer the question with which we are concerned in this paper.

(5) ***Arguments from various disabilities***

These arguments take the form, 'I grant you that you can make machines do all the things you have mentioned but you will never be able to make one to do X'. Numerous features X are suggested in this connexion. I offer a selection:

> Be kind, resourceful, beautiful, friendly, have initiative, have a sense of humour, tell right from wrong, make mistakes, fall in love, enjoy strawberries

and cream, make some one fall in love with it, learn from experience, use words properly, be the subject of its own thought, have as much diversity of behaviour as a man, do something really new.

No support is usually offered for these statements. I believe they are mostly founded on the principle of scientific induction. A man has seen thousands of machines in his lifetime. From what he sees of them he draws a number of general conclusions. They are ugly, each is designed for a very limited purpose, when required for a minutely different purpose they are useless, the variety of behaviour of any one of them is very small, etc., etc. Naturally he concludes that these are necessary properties of machines in general. Many of these limitations are associated with the very small storage capacity of most machines. (I am assuming that the idea of storage capacity is extended in some way to cover machines other than discrete-state machines. The exact definition does not matter as no mathematical accuracy is claimed in the present discussion.) A few years ago, when very little had been heard of digital computers, it was possible to elicit much incredulity concerning them, if one mentioned their properties without describing their construction. That was presumably due to a similar application of the principle of scientific induction. These applications of the principle are of course largely unconscious. When a burnt child fears the fire and shows that he fears it by avoiding it, I should say that he was applying scientific induction. (I could of course also describe his behaviour in many other ways.) The works and customs of mankind do not seem to be very suitable material to which to apply scientific induction. A very large part of space-time must be investigated, if reliable results are to be obtained. Otherwise we may (as most English children do) decide that everybody speaks English, and that it is silly to learn French.

There are, however, special remarks to be made about many of the disabilities that have been mentioned. The inability to enjoy strawberries and cream may have struck the reader as frivolous. Possibly a machine might be made to enjoy this delicious dish, but any attempt to make one do so would be idiotic. What is important about this disability is that it contributes to some of the other disabilities, *e.g.*, to the difficulty of the same kind of friendliness occurring between man and machine as between white man and white man, or between black man and black man.

The claim that 'machines cannot make mistakes' seems a curious one. One is tempted to retort, 'Are they any the worse for that?' But let us adopt a more sympathetic attitude, and try to see what is really meant. I think this criticism can be explained in terms of the imitation game. It is claimed that the interrogator could distinguish the machine from the man simply by setting them a number of problems in arithmetic. The machine would be unmasked because of its deadly accuracy. The reply to this is simple. The machine (programmed for playing the game) would not attempt to give the *right* answers to the arithmetic problems. It would deliberately introduce mistakes in a manner calculated to confuse the interrogator. A mechanical fault would probably show itself through an unsuitable decision as to what sort of a mistake to make in the arithmetic. Even this interpretation of the criticism is not sufficiently sympathetic.

But we cannot afford the space to go into it much further. It seems to me that this criticism depends on a confusion between two kinds of mistake. We may call them 'errors of functioning' and 'errors of conclusion'. Errors of functioning are due to some mechanical or electrical fault which causes the machine to behave otherwise than it was designed to do. In philosophical discussions one likes to ignore the possibility of such errors; one is therefore discussing 'abstract machines'. These abstract machines are mathematical fictions rather than physical objects. By definition they are incapable of errors of functioning. In this sense we can truly say that 'machines can never make mistakes'. Errors of conclusion can only arise when some meaning is attached to the output signals from the machine. The machine might, for instance, type out mathematical equations, or sentences in English. When a false proposition is typed we say that the machine has committed an error of conclusion. There is clearly no reason at all for saying that a machine cannot make this kind of mistake. It might do nothing but type out repeatedly '$0 = 1$'. To take a less perverse example, it might have some method for drawing conclusions by scientific induction. We must expect such a method to lead occasionally to erroneous results.

The claim that a machine cannot be the subject of its own thought can of course only be answered if it can be shown that the machine has *some* thought with *some* subject matter. Nevertheless, 'the subject matter of a machine's operations' does seem to mean something, at least to the people who deal with it. If, for instance, the machine was trying to find a solution of the equation $x^2 - 40x - 11 = 0$ one would be tempted to describe this equation as part of the machine's subject matter at that moment. In this sort of sense a machine undoubtedly can be its own subject matter. It may be used to help in making up its own programmes, or to predict the effect of alterations in its own structure. By observing the results of its own behaviour it can modify its own programmes so as to achieve some purpose more effectively. These are possibilities of the near future, rather than Utopian dreams.

The criticism that a machine cannot have much diversity of behaviour is just a way of saying that it cannot have much storage capacity. Until fairly recently a storage capacity of even a thousand digits was very rare.

The criticisms that we are considering here are often disguised forms of the argument from consciousness. Usually if one maintains that a machine *can* do one of these things, and describes the kind of method that the machine could use, one will not make much of an impression. It is thought that the method (whatever it may be for it must be mechanical) is really rather base. [...]

(6) Lady Lovelace's objection

Our most detailed information of Babbage's Analytical Engine comes from a memoir by Lady Lovelace. In it she states, 'The Analytical Engine has no pretensions to *originate* anything. It can do *whatever we know how to order it* to perform' (her italics). This statement is quoted by Hartree who adds: 'This does not imply that it may not be possible to construct electronic equipment which will "think for itself", or in which, in biological terms, one could set up a conditioned reflex, which would serve as a basis for "learning". Whether this is possible in principle or not is a stimulating and exciting question, suggested by some of these recent developments. But it did not seem that the machines constructed or projected at the time had this property'.

I am in thorough agreement with Hartree over this. It will be noticed that he does not assert that the machines in question had not got the property, but rather that the evidence available to Lady Lovelace did not encourage her to believe that they had it. It is quite possible that the machines in question had in a sense got this property. For suppose that some discrete-state machine has the property. The Analytical Engine was a universal digital computer, so that, if its storage capacity and speed were adequate, it could by suitable programming be made to mimic the machine in question. Probably this argument did not occur to the Countess or to Babbage. In any case there was no obligation on them to claim all that could be claimed.

This whole question will be considered again under the heading of learning machines.

A variant of Lady Lovelace's objection states that a machine can 'never do anything really new'. This may be parried for a moment with the saw, 'There is nothing new under the sun.' Who can be certain that 'original work' that he has done was not simply the growth of the seed planted in him by teaching, or the effect of following well-known general principles. A better variant of the objection says that a machine can never 'take us by surprise.' This statement is a more direct challenge and can be met directly. Machines take me by surprise with great frequency. This is largely because I do not do sufficient calculation to decide what to expect them to do, or rather because, although I do a calculation, I do it in a hurried, slipshod fashion, taking risks. Perhaps I say to myself, 'I suppose the voltage here ought to be the same as there: anyway let's assume it is.' Naturally I am often wrong, and the result is a surprise for me for by the time the experiment is done these assumptions have been forgotten. These admissions lay me open to lectures on the subject of my vicious ways, but do not throw any doubt on my credibility when I testify to the surprises I experience.

I do not expect this reply to silence my critic. He will probably say that such surprises are due to some creative mental act on my part, and reflect no credit on the machine. This leads us back to the argument from consciousness, and far from the idea of surprise. It is a line of argument we must consider closed, but it is perhaps worth remarking that the appreciation of something as surprising requires as much of a 'creative mental act' whether the surprising event originates from a man, a book, a machine or anything else.

The view that machines cannot give rise to surprises is due, I believe, to a fallacy to which philosophers and mathematicians are particularly subject. This is the assumption that as soon as a fact is presented to a mind all consequences of that fact spring into the mind simultaneously with it. It is a very useful assumption under many circumstances, but one too easily forgets that it is false. A natural consequence of doing so is that one then assumes that there is no virtue in the mere working out of consequences from data and general principles.

(7) ***Argument from continuity in the nervous system***

The nervous system is certainly not a discrete-state machine. A small error in the information about the size of a nervous impulse impinging on a neuron, may make a large difference to the size of the outgoing impulse. It may be argued that, this being so, one cannot expect to be able to mimic the behaviour of the nervous system with a discrete-state system.

It is true that a discrete-state machine must be different from a continuous machine. But if we adhere to the conditions of the imitation game, the interrogator will not be able to take any advantage of this difference. The situation can be made clearer if we consider some other simpler continuous machine. A differential analyser will do very well. (A differential analyser is a certain kind of machine not of the discrete-state type used for some kinds of calculation.) Some of these provide their answers in a typed form, and so are suitable for taking part in the game. It would not be possible for a digital computer to predict exactly what answers the differential analyser would give to a problem, but it would be quite capable of giving the right sort of answer. For instance, if asked to give the value of π (actually about 3·1416) it would be reasonable to choose at random between the values 3·12, 3·13, 3·14, 3·15, 3·16 with the probabilities of 0·05, 0·15, 0·55, 0·19, 0·06 (say). Under these circumstances it would be very difficult for the interrogator to distinguish the differential analyser from the digital computer.

(8) ***The argument from informality of behaviour***

It is not possible to produce a set of rules purporting to describe what a man should do in every conceivable set of circumstances. One might for instance have a rule that one is to stop when one sees a red traffic light, and to go if one sees a green one, but what if by some fault both appear together? One may perhaps decide that it is safest to stop. But some further difficulty may well arise from this decision later. To attempt to provide rules of conduct to cover every eventuality, even those arising from traffic lights, appears to be impossible. With all this I agree.

From this it is argued that we cannot be machines. I shall try to reproduce the argument, but I fear I shall hardly do it justice. It seems to run something like this. 'If each man had a definite set of rules of conduct by which he regulated his life he would be no better than a machine. But there are no such rules, so men cannot be machines.' The undistributed middle is glaring. I do not think the argument is ever put quite like this, but I believe this is the argument used nevertheless. There may however be a certain confusion between 'rules of conduct' and 'laws of behaviour' to cloud the issue. By 'rules of conduct' I mean precepts such as 'Stop if you see red lights', on which one can act, and of which one can be conscious. By 'laws of behaviour' I mean laws of nature as applied to a man's body such as 'if you pinch him he will squeak'. If we substitute 'laws of behaviour which regulate his life' for 'laws of conduct by which he regulates his life' in the argument quoted the undistributed middle is no longer insuperable. For we believe that it is not only true that being regulated by laws of behaviour implies being some sort of machine (though not necessarily a discrete-state machine), but that conversely being such a machine implies being regulated by such laws. However, we cannot so easily convince ourselves of the absence of complete laws of behaviour as of complete rules of conduct. The only way we know of for finding such laws is scientific observation, and we certainly know of no circumstances under which we could say, 'We have searched enough. There are no such laws.'

We can demonstrate more forcibly that any such statement would be unjustified. For suppose we could be sure of finding such laws if they existed. Then given a discrete-state machine it should certainly be possible to discover by observation sufficent about it to predict its future behaviour, and this within a reasonable time, say a thousand years. But this does not seem to be the case. I have set up on the Manchester computer a small program using only 1000 units of storage, whereby the machine supplied with one sixteen figure number replies with another within two seconds. I would defy anyone to learn from these replies sufficient about the program to be able to predict any replies to untried values.

(9) The argument from extra-sensory perception

I assume that the reader is familiar with the idea of extra-sensory perception, and the meaning of the four items of it, viz. telepathy, clairvoyance, precognition and psycho-kinesis. These disturbing phenomena seem to deny all our usual scientific ideas. How we should like to discredit them! Unfortunately the statistical evidence, at least for telepathy, is overwhelming. It is very difficult to rearrange one's ideas so as to fit these new facts in. Once one has accepted them it does not seem a very big step to believe in ghosts and bogies. The idea that our bodies move simply according to the known laws of physics, together with some others not yet discovered but somewhat similar, would be one of the first to go.

This argument is to my mind quite a strong one. One can say in reply that many scientific theories seem to remain workable in practice, in spite of clashing with E.S.P.; that in fact one can get along very nicely if one forgets about it. This is rather cold comfort, and one fears that thinking is just the kind of phenomenon where E.S.P. may be especially relevant.

A more specific argument based on E.S.P. might run as follows: 'Let us play the imitation game, using as witnesses a man who is good as a telepathic receiver, and a digital computer. The interrogator can ask such questions as "What suit does the card in my right hand belong to?" The man by telepathy or clairvoyance gives the right answer 130 times out of 400 cards. The machine can only guess at random, and perhaps gets 104 right, so the interrogator makes the right identification.' There is an interesting possibility which opens here. Suppose the digital computer contains a random number generator. Then it will be natural to use this to decide what answer to give. But then the random number generator will be subject to the psycho-kinetic powers of the interrogator. Perhaps this psycho-kinesis might cause the machine to guess right more often than would be expected on a probability calculation, so that the interrogator might still be unable to make the right identification. On the other hand, he might be able to guess right without any questioning, by clairvoyance. With E.S.P. anything may happen.

If telepathy is admitted it will be necessary to tighten our test up. The situation could be regarded as analogous to that which would occur if the interrogator were talking to himself and one of the competitors was listening with his ear to the wall. To put the competitors into a 'telepathy-proof room' would satisfy all requirements.

Learning machines

The reader will have anticipated that I have no very convincing arguments of a positive nature to support my views. If I had I should not have taken such pains to point out the fallacies in contrary views. Such evidence as I have I shall now give.

Let us return for a moment to Lady Lovelace's objection, which stated that the machine can only do what we tell it to do. One could say that a man can 'inject' an idea into the machine, and that it will respond to a certain extent and then drop into quiescence, like a piano string struck by a hammer. Another simile would be an atomic pile of less than critical size: an injected idea is to correspond to a neutron entering the pile from without. Each such neutron will cause a certain disturbance which eventually dies away. If, however, the size of the pile is sufficiently increased, the disturbance caused by such an incoming neutron will very likely go on and on increasing until the whole pile is destroyed. Is there a corresponding phenomenon for minds, and is there one for machines? There does seem to be one for the human mind. The majority of them seem to be 'sub-critical,' *i.e.* to correspond in this analogy to piles of sub-critical size. An idea presented to such a mind will on average give rise to less than one idea in reply. A smallish proportion are super-critical. An idea presented to such a mind may give rise to a whole 'theory' consisting of secondary, tertiary and more remote ideas. Animals minds seem to be very definitely sub-critical. Adhering to this analogy we ask, 'Can a machine be made to be super-critical?'

The 'skin of an onion' analogy is also helpful. In considering the functions of the mind or the brain we find certain operations which we can explain in purely mechanical terms. This we say does not correspond to the real mind: it is a sort of skin which we must strip off if we are to find the real mind. But then in what remains we find a further skin to be stripped off, and so on. Proceeding in this way do we ever come to the 'real' mind, or do we eventually come to the skin which has nothing in it? In the latter case the whole mind is mechanical. (It would not be a discrete-state machine however. We have discussed this.)

These last two paragraphs do not claim to be convincing arguments. They should rather be described as 'recitations tending to produce belief.'

The only really satisfactory support that can be given for the view expressed [*above*], will be that provided by waiting for the end of the century and then doing the experiment described. But what can we say in the meantime? What steps should be taken now if the experiment is to be successful?

As I have explained, the problem is mainly one of programming. Advances in engineering will have to be made too, but it seems unlikely that these will not be adequate for the requirements. Estimates of the storage capacity of the brain vary from 10^{10} to 10^{15} binary digits. I incline to the lower values and believe that only a very small fraction is used for the higher types of thinking. Most of it is probably used for the retention of visual impressions. I should be surprised if more than 10^9 was required for satisfactory playing of the imitation game, at any rate against a blind man. (Note – The capacity of the *Encyclopaedia Britannica*, 11th edition, is 2×10^9.) A storage capacity of 10^7 would be a very practicable possibility even by present techniques. It is probably not necessary to increase the speed of operations of the machines at all. Parts of modern machines which can be regarded as analogues of nerve cells work about a thousand times faster than the latter. This should provide a 'margin of safety' which could cover losses of speed arising in many ways. Our problem then is to find out how to program these machines to play the game. At my present rate of working I produce about a thousand digits of program a day, so that about sixty workers, working steadily through the fifty years might accomplish the job, if nothing went into the waste-paper basket. Some more expeditious method seems desirable.

In the process of trying to imitate an adult human mind we are bound to think a good deal about the process which has brought it to the state that it is in. We may notice three components,

(a) The initial state of the mind, say at birth,
(b) The education to which it has been subjected,
(c) Other experience, not to be described as education, to which it has been subjected.

Instead of trying to produce a program to simulate the adult mind, why not rather try to produce one which simulates the child's? If this were then subjected to an appropriate course of education one would obtain the adult brain. Presumably the child-brain is something like a note-book as one buys it from the stationers. Rather little mechanism, and lots of blank sheets. (Mechanism and writing are from our point of view almost synonymous.) Our hope is that there is so little mechanism in the child-brain that something like it can be easily programmed. The amount of work in the education we can assume, as a first approximation, to be much the same as for the human child.

We have thus divided our problem into two parts. The child-program and the education process. These two remain very closely connected. We cannot expect to find a good child machine at the first attempt. One must experiment with teaching one such machine and see how well it learns. One can then try another and see if it is better or worse. There is an obvious connection between this process and evolution, by the identifications

Structure of the child machine	=	Hereditary material
Changes of the child machine	=	Mutations
Natural selection	=	Judgment of the experimenter

One may hope, however, that this process will be more expeditious than evolution. The survival of the fittest is a slow method for measuring advantages. The experimenter, by the exercise of intelligence, should be able to speed it up. Equally important is the fact that he is not restricted to random mutations. If he can trace a cause for some weakness he can probably think of the kind of mutation which will improve it.

It will not be possible to apply exactly the same teaching process to the machine as to a normal child. It will not, for instance, be provided with legs, so that it could not be asked to go out and fill the coal scuttle. Possibly it might not have eyes. But however well these deficiencies might be overcome by clever engineering, one could not send the creature to school without the other children making excessive fun of it. It must be given some tuition. We need not be too concerned about the legs, eyes, etc. The example of Miss Helen Keller shows that education can take place provided that communication in both directions between teacher and pupil can take place by some means or other.

We normally associate punishments and rewards with the teaching process. Some simple child-machines can be constructed or programmed on this sort of principle. The machine has to be so constructed that events which shortly preceded the occurrence of a punishment-signal are unlikely to be repeated, whereas a reward-signal increased the probability of repetition of the events which led up to it. These definitions do not presuppose any feelings on the part of the machine. I have done some experiments with one such child-machine, and succeeded in teaching it a few things, but the teaching method was too unorthodox for the experiment to be considered really successful.

The use of punishments and rewards can at best be a part of the teaching process. Roughly speaking, if the teacher has no other means of communicating to the pupil, the amount of information which can reach him does not exceed the total number of rewards and punishments applied. By the time a child has learnt to repeat 'Casabianca' he would probably feel very sore indeed, if the text could only be discovered by a 'Twenty Questions' technique, every 'NO' taking the form of a blow. It is necessary therefore to have some other 'unemotional' channels of communication. If these are available it is possible to teach a machine by punishments and rewards to obey orders given in some language, *e.g.* a symbolic language. These orders are to be tramsmitted through the 'unemotional' channels. The use of this language will diminish greatly the number of punishments and rewards required.

Opinions may vary as to the complexity which is suitable in the child machine. One might try to make it as simple as possible consistently with the general principles. Alternatively one might have a complete system of logical inference 'built in'. In the latter case the store would be largely occupied with definitions and propositions. The propositions would have various kinds of status, *e.g.* well-established facts, conjectures, mathematically proved theorems, statements given by an authority, expressions having the logical form of proposition but not belief value. Certain propositions may be described as 'imperatives'. The machine should be so constructed that as soon as an imperative is classed as 'well-established' the appropriate action automatically takes place. To illustrate this, suppose the teacher says to the machine, 'Do your homework now'. This may cause 'Teacher says "Do your homework now"' to be included amongst the well-established facts. Another such fact might be, 'Everything that teacher says is true'. Combining these may eventually lead to the imperative, 'Do your homework now,' being included amongst the well-established facts, and this, by the construction of the machine, will mean that the homework actually gets started, but the effect is very satisfactory. The processes of inference used by the machine need not be such as would satisfy the most exacting logicians. There might for instance be no hierarchy of types. But this need not mean that type fallacies will occur, any more than we are bound to fall over unfenced cliffs. Suitable imperatives (expressed *within* the systems, not forming part of the rules *of* the system) such as 'Do not use a class unless it is a subclass of one which has been mentioned by teacher' can have a similar effect to 'Do not go too near the edge'.

The imperatives that can be obeyed by a machine that has no limbs are bound to be of a rather intellectual character, as in the example (doing homework) given above. Important amongst such imperatives will be ones which regulate the order in which the rules of the logical system concerned are to be applied. For at each stage when one is using a logical system, there is a very large number of alternative steps, any of which one is permitted to apply, so far as obedience to the rules of the logical system is concerned. These choices make the difference between a brilliant and a footling reasoner, not the difference between a sound and a fallacious one. Propositions leading to imperatives of this kind might be 'When Socrates is mentioned, use the syllogism in Barbara' or 'If one method has been proved to be quicker than another, do not use the slower method.' Some of these may be 'given by authority', but others may be produced by the machine itself, *e.g.* by scientific induction.

The idea of a learning machine may appear paradoxical to some readers. How can the rules of operation of the machine change? They should describe completely how the machine will react whatever its history might be, whatever changes it might undergo. The rules are thus quite time-invariant. This is quite true. The explanation of the paradox is that the rules which get changed in the learning process are of a rather less pretentious kind, claiming only an ephemeral validity. The reader may draw a parallel with the Constitution of the United States.

An important feature of a learning machine is that its teacher will often be very largely ignorant of quite what is going on inside, although he may still be able to some extent to predict his pupil's behaviour. This should apply most strongly to the later education of a machine arising from a child-machine of well-tried design (or program). This is in clear contrast with normal procedure when using a machine to do computations: one's object is then to have a clear mental picture of the state of the machine at each moment in the computation. This object can only be achieved with a struggle. The view that 'the machine can only do what we know how to order it to do', appears strange in face of this. Most of the programs which we can put into the machine will result in its doing something that we cannot make sense of at all, or which we regard as completely random behaviour. Intelligent behaviour presumably consists in a departure from the completely disciplined behaviour involved in computation, but a rather slight one, which does not give rise to random behaviour, or to pointless repetitive loops. Another important result of preparing our machine for its part in the imitation game by a process of teaching and learning is that 'human fallibility' is likely to be omitted in a rather natural way, *i.e.* without special 'coaching.' [...] Processes that are learnt do not produce a hundred per cent certainty of result; if they did they could not be unlearnt.

It is probably wise to include a random element in a learning machine. A random element is rather useful when we are searching for a solution of some problem. Suppose for instance we wanted to find a number between 50 and 200 which was equal to the square of the sum of its digits, we might start at 51 then try 52 and go on until we got a number that worked. Alternatively we might choose numbers at random until we got a good one. This method has the advantage that it is unnecessary to keep track of the values that have been tried, but the disadvantage that one may try the same one twice, but this is not very important if there are several solutions. The systematic method has the disadvantage that there may be an enormous block without any solutions in the region which has to be investigated first. Now the learning process may be regarded as a search for a form of behaviour which will satisfy the teacher (or some other criterion). Since there is probably a very large number of satisfactory solutions the random method seems to be better than the systematic. It should be noticed that it is used in the analogous process of evolution. But there the systematic method is not possible. How could one keep track of the different genetical combinations that had been tried, so as to avoid trying them again?

We may hope that machines will eventually compete with men in all purely intellectual fields. But which are the best ones to start with? Even this is a difficult decision. Many people think that a very abstract activity, like the playing of chess, would be best. It can also be maintained that it is best to provide the machine with the best sense organs that money can buy, and then teach it to understand and speak English. This process could follow the normal teaching of a child. Things would be pointed out and named, etc. Again I do not know what the right answer is, but I think both approaches should be tried.

We can only see a short distance ahead, but we can see plenty there that needs to be done.

Social implications of intelligent machines

Margaret A. Boden

Current achievements and future developments

Computer hardware gets steadily cheaper: it is predicted that by the year 2000, a 65k bit silicon chip capable of 20 million instructions a second will sell for one US dollar. But the social application of intelligent machines also demands advances in software, such as more powerful programming languages and improved organization and use of knowledge. Work in artificial intelligence has shown that the problems of organizing and accessing large data bases will not be quickly solved. The early optimism in the field has waned accordingly: already by 1962, the mediocrity of chess programs was being stressed by someone who in 1957 had predicted that a computer would be world chess champion within 10 years. And much current research aims to develop organizational principles whereby a knowledge-domain can be economically represented and appropriately addressed without triggering a combinatorial explosion.

Nevertheless, professionals participating in a multistage 'Delphi' forecasting exercise have predicted that within the next 30 years (in many cases, within only 10 or 15 years), social applications of artificial intelligence will be widely, i.e. commercially, available. In general, the prototype is expected 5 $\pm$ 2 years ahead of the commercial version. The dates I shall mention are taken from this Delphi study (although my own view is that these predictions tend to underestimate the difficulties involved).

The applications forecast run from robot housecleaners, chauffeurs, and industrial workers, through programmed gameplayers and storytellers, to automatic teachers, physicians, legal justices, marriage counsellors, and literary critics. In all these cases, the emphasis is on reasoned and flexible judgment on the program's part, as opposed to the storage and regurgitation of isolated facts, or the repetitive performance of a fixed sequence of discriminations and movements.

For example, the computer diagnosticians of the 1980s will not simply store lists of symptom-diagnosis pairs, or prescribe treatment in a blindly dogmatic (and apparently 'objective') fashion. One current prototype is the MYCIN system, an interactive program that simulates a medical consultant specializing in infectious diseases. It engages in question-and-answer conversations (lasting 20 minutes on average) with doctors needing specialist help, and in 75 per cent of cases gives the same counsel as a human expert. The physician asks MYCIN for advice on the identification of micro-organisms and the prescription of antibiotics, and also for explanations of its advice expressed at the appropriate level of detail.

MYCIN's explanatory capacity enables physicians who disagree with specific aspects of the program's clinical rationale rationally to *reject* MYCIN's advice. It also helps non-specialist doctors to learn more about the complexities of diagnosis and therapy in this class of diseases. And it allows human consultants to make general improvements in the program, by telling it about relevant knowledge that they realize in specific cases to be missing or inadequately stated.

A program like MYCIN involves artificial intelligence techniques. Quite apart from its (rather restricted) natural language understanding, its ability to explain itself on many different levels of detail implies a self-knowledge of its reasoning and goal-structure that are crucial to intelligent thinking. And its ability to learn by being told implies some mastery of the problem of making spontaneous inferences on the basis of input information, though admittedly only for a very limited area of discourse.

Many features of MYCIN would be embodied also in programmed legal arbiters (prototype predicted by 1988). These would not only search for relevant legal precedents in the judicial literature – a far from trivial task – but would also offer legal advice. Like MYCIN, legal arbiters will preferably be used to *augment* human judgment, rather than *replace* it. Accordingly, like MYCIN, they should when appropriate offer several (reasoned) alternative judgments, not just the one of which they are most confident. MYCIN's assessment of degrees of confidence is not a mere statistical probability measure. It takes into account psychological factors about the evidential relations of beliefs, factors that philosophers of science have considered in regard to 'confirmation theory'. Legal programs, too, would have to incorporate more subtle concepts of *confidence* and *evidence* than Bayesian probability, in order to avoid judicial absurdities of various types.

Flexible planning, and an intelligently structured representation of knowledge, is used by the Computer-Based Consultant, a system designed to give on-the-job advice about how to assemble a machine to novice mechanics having varied levels of expertise. The program uses the specific queries posed by the human novice as cues directing it to answer at one level or another. Thus it tells one mechanic simply to 'Replace the pump', but advises a less experienced person first to 'Remove the 4 mounting bolts at the base of the pump using a 3/8-inch open-end wrench'. When a human failure occurs because of an unexpected happening (including but not restricted to previous mistakes on the novice's part), the Consultant can question the mechanic in an intelligent fashion so as to locate the difficulty. In most cases, this will not be a simple matter of asking 'What's the trouble?' since the novice usually does not know just what has gone wrong, and may claim to have followed the program's advice to the letter. Because the program's representation of the semantics of this domain includes detailed and intelligently structured knowledge, in the majority of cases it is able to spot the trouble and work out a way of putting the human worker back on the right track.

Future automatic tutors will be more sophisticated than today's computer-aided instruction systems, even those that allow differential branching of the 'syllabus' according to the student's mistakes and queries. Like MYCIN and the Computer-Based-Consultant, they will be able to initiate and answer questions at various levels of detail, according to the pupil's range of expertise. They will concentrate on asking probing questions enabling them to model the student's understanding of the topic, and will devise an individually tailored tutorial strategy designed to build on this understanding in fruitful ways. Such programs presuppose an intelligently structured representation of the knowledge concerned, in its various aspects and degrees. Commercial availability is forecast for 1988, and exploratory programs already exist

Public access to a powerful computer would be required for many of the predicted social applications of artificial intelligence. John McCarthy accordingly has forecast the widespread use of video-teletype *home information terminals,* linked to a national network of time-sharing computers.

A speech-understanding program would allow the user to speak requests instead of typing them. At present, this can be reliably done (for instance, by the HEARSAY system for playing voice-chess) only by using a very small and deliberately distinct vocabulary, with unnatural pauses between the words. This is because (as in the interpretation of visual scenes) the continuous stream of normal speech has to be sensibly segmented into individual words before it can be understood – but, as artificial intelligence research has shown, it has first to be understood in order to be segmented. Even the individual sounds cannot be distinguished on purely auditory grounds, but only by reference to the wider linguistic context. However, commercial uses are forecast for 1983.

According to some forecasts, the stationary home terminal will in 1995 be supplemented by a domestic robot. It should not be necessary to clear the floor so that the robot can vacuum it: the robot itself will supposedly be able to recognize the objects littering the room and put them in their proper place – or, by default, in one corner. Current achievements suggest that a robot might fairly soon be able to recognize a waste-paper bin wrongly placed by the window, or even high up on the table, so as to reposition it by the desk. But *only* if the room is otherwise unnaturally tidy, and *only* if the bin, window, table, and desk are of uniform type.

Industrial and agricultural robots (forecast by 1980) are less fanciful. These would not have a fixed sensori-motor capability, like the automated machines of today. Instead, they would be flexible in operation, and could learn *new* tasks. They would learn not by being laboriously reprogrammed, but by being shown examples of (for instance) new machine parts and machine tools, and by being given an outline sketch of the desired procedure which they would then elaborate in detail. The sort of planning facilities used by the Computer-Based-Consultant, or programs like BUILD, would be crucial here. BUILD plans how to assemble brick-structures which may require 'creative' steps such as using a temporary scaffold, counterweight, or support. Features like potential stability and steadiness of movement are continually taken into account in deciding what to do, and if a safer method can be found of constructing the desired building then that method will be used. If it turns out that the 'safer' method is not safe after all, BUILD can alter course accordingly, meanwhile losing none of the information gained during the aborted attempt. Slight difficulties can be recognized and overcome without affecting the general strategy, but graver obstacles prompt radical replanning.

There are already some mobile carrier-robots, and programmed hand–eye systems that can assemble simple machines from components they learn to recognize by being shown examples or that (like MIT's COPY-DEMO) can build visually demonstrated structures out of a 'warehouse' of familiar parts. Moreover, a robot linked to the Computer-Based-Consultant would be able to assemble a pump even if someone else had left the job half-finished, or had wrongly positioned some of the parts. This flexible behaviour is very different from that of current industry's computer-controlled devices, such as the Unimate, which require everything to be in the right place at the right time.

Possible effects on concepts of self and society

The potential social influence of artificial intelligence is ambiguous, for two reasons. First, specific applications (as of any science) may be used for good or evil ends, and may have unsuspected side-effects. Second, much depends on the background human context, including the general public's implicit philosophical assumptions or 'image of man'. I shall first discuss some possible bad effects, and then sketch potential good effects of intelligent machines.

There is a widespread suspicion of artificial intelligence among the general public, based on the common philosophical assumption that regarding man as a computational mechanism entails denying human subjectivity, individuality, and moral freedom. This suspicion has been voiced by critics of urban industrialism such as Theodore Roszak and Herbert Marcuse, and by psychologists in touch with laymen seeking practical help. For example, this complaint is from Rollo May, a therapeutic and counselling psychologist:

> I take very seriously ... the dehumanizing dangers in our tendency in modern science to make man over into the image of the machine, into the image of the techniques by which we study him. ... A central core of modern man's 'neurosis' is the undermining of his experience of himself as responsible, the sapping of his willing and decision.

May's point is that choices made without confidence in their possible relevance to eventual action are unlikely to be effective – or perhaps even to be made at all. For if a man's self-image represents himself to himself as an autonomous purposive creature capable of pursuing certain ends, then it can be used to generate choices and guide his action accordingly. Even in machines, as is evident in planning programs such as BUILD, the internal representation of the possible modes of action that are available to the system can be crucial in directing performance. But if a 'depersonalization' of the self image occurs so that the self is no longer seen as a truly purposive system, then relatively *in*human, 'pathological' behaviour can be expected in consequence.

This type of degenerate self-model is encouraged by artificial intelligence in general, given the popular (though mistaken) philosophical assumption of total incompatibility between mechanism and humanism. What May calls 'the undermining of one's experience of oneself as responsible' may therefore be exacerbated by the development of clever programs.

As well as having immediate consequences in one's personal life, this sapping of willpower can have widespread social implications. For example, the political institution of participatory democracy assumes an ascription of responsibility to individuals which fits ill with the dehumanized image remarked by May. Consequently, providing citizens with home terminals, with the partial aim of enabling them to vote and express their political views without leaving their fireside, might subtly undermine their sense of civic responsibility. This is the reverse of the intended effect, which is to encourage individual citizens to engage more fully in democratic government, thereby lessening the common feeling of alienated helplessness with respect to the governmental process. (Of course, to *feel* less a cog is not necessarily to *be* any less a cog: the home terminal might function as a subtle form of social control, damping down dissent by contenting people with an illusory sense of political participation.)

While any intelligent program may have a dehumanizing effect on people who see an unbridgeable metaphysical gulf between themselves and machines, as many people do, some would be especially open to this criticism. for instance, the prediction that by the year 2000 automatic interviewers will be used to aid the diagnosis and treatment of psychiatric patients is a suggestion that many would spurn. Thus Joseph Weizenbaum, the creator of the early conversational program ELIZA, has bitterly denounced the 'obscene' idea of employing programs in clinical situations.

In medical contexts dealing with basically physical illness, artificial intelligence may be more welcome. The MYCIN system has not yet been tried out in a clinical situation, though the authors report some resistance on the part of clinicians to the idea of using the program. When MYCIN has achieved a 90 per cent match with human experts, it will be introduced experimentally into a hospital; only then will its authors be able to see who uses it, how often, and what effect it has on the prescribing practices of doctors and the clinical status of their patients. But a much simpler program for the diagnosis of peptic ulcers has been used on patients, who often claim to prefer this diagnostic method. Ironically, they describe the machine (with which they communicate by teletype) as more friendly, polite, relaxing, and comprehensible than the average physician.

These chastening observations about the superiority of the personal habits of programs over those of human doctors can doubtless be expected also with regard to automatic lawyers, bureaucrats, and teachers. While perhaps appreciated in isolated interactions of a tedious, technical, or embarrassing nature, this imperturbable mode might come to be consciously spurned in human relations in general, with a consequent increased emphasis on emotional spontaneity. Alternatively, the opposite effect might result: the blandness of one's guest-computer might come to be emulated in one's own mode of expression – or that of one's children. (The possible ill effect on children was cited in the Delphi survey as a disadvantage of the domestic robot.)

In general, one has to consider the dehumanizing effects of people's becoming decreasingly dependent upon human contact for satisfaction of their needs. Many who today can do their jobs only by going to a particular place of work, might tomorrow be able to stay at home and communicate with their clients and co-workers via the home terminal. The socially isolating influence of television is as nothing to the alienation and loneliness that might result from over-enthusiastic reliance on the home terminal and associated gadgetry.

Let us now ask whether social applications of artificial intelligence might have any *welcome* effects on the way people view themselves and other people.

Computational models of intelligence are in fact markedly more human than the behaviourist models of mankind that have been widely accepted for years, because the computational approach can endorse the humanist's stress on the idiosyncracy of people's subjective world-views and on the directive role of the self-image. A program constructs its interpretation of the input by way of its particular epistemological scheme, or set of inner models of the world, and the same input may thus be 'experienced' very differently by different programs. If this commonly unsuspected 'humanizing' feature of artificial intelligence can be brought home to the general public, then many of the ill effects I have hypothesized will be allayed.

Educational methods based on the pedagogical philosophy of LOGO-turtles might change ways of thinking about 'failure'. Instead of the passively defeatist 'I'm not good at this', the child would say 'How can I make myself better at it?' This attitude is encouraged by the computational way of thinking about thinking, with its emphasis on the creative interrelation of many different procedures, and on the unintended effects of specifiable bugs in basically well-conceived attempts to achieve one's goal. By contrast, constructive self-criticism is not encouraged by a conception of intelligence that views it as the product of a number of mysterious monolithic 'talents' or abilities. which one either has or lacks, willy-nilly. (An increase in the use of programming in schools may also help to avert the growth of a socially divisive 'computer elite', a small group of people whose members are the only ones to understand computation and so the only ones not to feel alienated in the computerized society of the future.)

Publicly available programs are unlikely to be rendered 'emotional', even assuming this to be in principle possible, since there would be little point in doing so. If people still felt a need to draw a line between themselves and 'machines', their valuation of the emotional life would probably increase, with corresponding effects on cultural mores. Incorporated in the self-image of most Westerners is the Protestant Ethic that only hard work is a really serious activity. And 'work' is implicitly defined as paid, as done in one's employer's time rather than in one's own 'leisure' time, and at the employer's behest rather than for one's own purposes. Consequently, massive unemployment could be more soul-destroying than the most repetitive of factory jobs, causing men destructively to see themselves as social parasites. This is less likely to happen to women, who enjoy an internalized acceptance of emotional values that enables then to derive greater fulfilment from personal relations and the expression of emotions outside the immediate family circle. If automation increases men's opportunities for human interaction, with friends as well as family, we may expect radical changes in the social definitions of sexual roles (for emotionality at present is seen as secondary to the masculine role). These changes would be due primarily to economic shifts in working-hours and the sexual division of labour, but they could be reinforced by a general increase in evaluation of the emotional life deriving from the 'emotionless' nature of programs.

Precautionary measures in writing and presenting programs

Sometimes things will go wrong, so that a program needs to be adjusted. Programmers who wish to know what is going on, and what needs to be stopped should anything go amiss, must take steps beforehand to allow for this. Their programs should be intelligible and explicit, so that 'what is going on' is not buried in the code or implicitly embodied in procedures whose aim and effect is obscure.

Programs should be generously commented, so that what a procedure is *supposed* to be doing (and *why*) is readily visible. The importance of this for aiding debugging (whether by a human or an automatic programmer) is evident from the program-writing program, HACKER. HACKER writes programs to perform tasks such as stacking bricks in specified ways, and is able to correct its own mistakes – and to avoid similar mistakes in future tasks of a generally similar nature – by way of its understanding of the purposive structure of task and program alike. HACKER uses the intention-coding comments it attaches to each line of its 'first draft' programs in amending these programs later. Only because it has such a good idea of what it is trying to do, and how it is trying to do it, is it able self-critically to modify its own procedures.

Intelligibility and explicitness are to some extent opposed, and programs get less readable as they approach the machine code level. What counts as 'machine code' is likely to become rather more intelligible to human beings, since instructions that now have to be programmed may be 'hardwired' into the electronics of the machine.

To economize on computation time, the next best thing to hardwiring is compiling. It would in general be advantageous to have the possibility of switching from compiled to interpreted mode if necessary, so as intelligently to guide the giving of detailed instructions to the machine in light of current circumstances, and the programmer should specify the sorts of contexts in which this switch might be advisable. The learning program HACKER, when in doubt (and when running a new program for the first time), can switch into a slower 'careful' mode in which every step is carefully examined before it is taken.

It is easier to see what is going on if subroutines are written so that they can easily be 'got out' from the program as a whole. This principle of modular programming is exemplified in programs like BUILD. The understanding and improvement of BUILD are facilitated by the clear distinction between the intercommunicating 'expert' subroutines. Modular programming will be required also in writing programs that cannot possibly be spied on or maliciously altered. For unless it is small, a program cannot be proved to perform *exactly* the functions required by the designer *and no other functions whatever.* The only way to make unauthorized access absolutely unfeasible would be to build the system around individually proved modules, or 'security kernels'. Modular programming also helps to counter the conservatism inherent in widespread applications of very complex systems, for if faults can be adjusted without necessitating widespread tinkering with the system then programmers will be better able to face the task of improving a large-scale system.

If programmers are to be able to see what is going on, they should not use programming techniques which – while making for readable programs – render the control structure obscure. For example, use of the programming language PLANNER tends to produce readable programs in which one cannot tell what is going on by examining the code. PLANNER programs are legible because it is a 'goal-directed' language in which routines can be indirectly invoked by way of general 'goal-patterns' matching the index of the routines in question. Specific advice to try one routine before another can be included by the programmer. But the language embodies an automatic backtrack facility which tags on to the end of the specific advice (if any) the instruction to USE ANYTHING that might work: that is, any routine whose index-pattern matches the goal currently being sought. This strategy might lead to some nasty situations: a distraught parent in squalid surroundings may try anything to stop the baby screaming – and succeed appallingly well. A human has to be distraught to disregard the side-effects of 'effective' measures, but a program may not even know about them.

Moreover, in the PLANNER automatic backtrack situation, the program simply tries out the first method that seems apt, since it has no way of comparing all possibilities beforehand. The knowledge that hitting may hurt the baby could be buried inside the relevant procedure, to be found only when this way of making the baby quiet was run. Only if the programmer had specifically included the advice never to hit the baby, would a PLANNER program refuse even to consider it. The programming language CONNIVER, (which was developed largely in response to the difficulties involved in PLANNER backtracking), by contrast does allow for the potential choices to be listed for higher-level criticism (BUILD uses this facility in deciding on the safest way to build a brick-house). If a CONNIVER program knew that hitting hurts, and was able to access this knowledge when needed, it would be able to control itself long enough to find an alternative way of quieting the baby, or to decide to abandon this goal as not legally achievable.

This example makes it clear that a flexibly intelligent control structure is only useful if the program embodies sensible criteria of what effects are 'undesirable'. Some 'sensible' criteria are culture-specific, and one may expect much moral-political disagreement about what precautions regarding artificial intelligence are worthwhile. Isaac Asimov's well-known 'Three Laws of Robotics' each assume that we know (and have communicated to the robot's program) what is to be counted as 'harm'.

Should we, for instance, regard it as harmful to lie? (We commonly say it is immoral, and utilitarians argue that it is socially harmful.) But if so, we may be landed with a teletyping program that is forbidden to tell tactful white lies. One of the reasons people often have for lying is precisely to *avoid* harming others. The more 'personal' the program, the more likely that lying (or a tell-tale silence?) might be in order. Some people would even claim that for social-political reasons – such as preventing dissent, disorder, or panic – certain 'impersonal facts' should be kept from the general public. On this view, should all home terminal programs (particularly those notionally contributing to participatory government) be kept ignorant of them also, or should they be able intelligently to take account of them while 'protectively' keeping them to themselves, if necessary lying in order to do so?

The ethical ambiguity of lying is only one of many difficult cases. One may therefore experience some reserve about recommendations of a new profession of value-impact forecasters: ethical experts armed with scientific tools for making cost-benefit judgments. Roszak has sourly commented: 'What these ethical engineers will know of "value" (Old Style: the meaning of life) may of course be only a computer simulation of a statistical illusion gleaned from questionnaires whose unreality crudely approximates a moral imbecile's conception of an ethical decision.'

Roszak regards the use of psychological terms in psychological contexts, and artificial intelligence research in general, as inescapably dehumanizing for the culture that admits them. He would urge programmers not merely to refuse to write obviously exploitative programs, but to re-examine the philosophical assumptions of their whole enterprise. In view of the discussion above, this surely is not too much to ask. Psychological theories in general are not purely *descriptive,* but are largely *constitutive* of social reality, and computational theories are no exception. If the public believes – rightly or wrongly – that science regards people as 'nothing but clockwork', then clockwork-people we may tend to become. This is why computer scientists should stress the basic philosophical compatibility of 'humanist' and computational views.

In this connection, perhaps programs for public use should include explicit reminders of some of the differences between computers and people. This is done (for utilitarian reasons) in the peptic ulcer program previously described: it continually reminds patients that they are on-line to a machine. The reminder helps to avoid mystification of the patients, stressing that they are merely 'filling in a form' by teletype rather than engaging in flexible communication, still less participating in a human relationship. 'Plausibility-tricks' are sometimes included in language-using programs. For instance, the CAI programs currently used in schools take care to address the children from time to time by their first names, so as to put them at their ease. In view of the dehumanizing potential of computer applications (quite apart from the possible results of an overgenerous misunderstanding by the person), it may be that the limits of individual programs should be made as clear as possible, to users, and plausibility tricks used sparingly – if at all.

On one point, however, even Roszak presumably would agree that the *similarity* between people and programs should be deliberately stressed. One of his objections to the social use of computers in an advisory capacity is that the machine may be seen by the public (including the politicians) as purely *objective,* and therefore not to be argued with. Accordingly, the inescapable subjectivity of a program's judgments should be made clear to its users. We saw that this is done implicitly in the MYCIN medical consultant: MYCIN not only offers alternative judgments when it perceives several possibilities of diagnosis or treatment, but gives its reasons for each so that physicians can rationally reject its advice if they see fit. In general, it would be worthwhile somehow to remind the user that programs function within their own subjective cognitive worlds, just as we do. Any differences between the two are matter for epistemological debate, not for servile capitulation on the part of the person.

Conclusions

Increasingly intelligent machines are likely to become socially available, whether for use by the general public or by political and administrative institutions.

Many of the potential ill-effects depend on the common (though mistaken) view that mechanism is incompatible with 'human' qualities. If the public assume that science offers an image of man that is irreconcilable with humanism, they must either deny their humanity – with socially destructive results – or else forfeit a scientific understanding of mankind.

So in addition to writing programs in a sensible fashion (so that the control structure is perspicuous and alteration is facilitated), professionals involved in artificial intelligence should take pains to see that this choice is not regarded by the public as unavoidable. For example, they should avoid 'plausibility tricks' in programs that might lead to unnecessary mystification of the users, and they should point out that a program's data and inference-procedures may always be questioned in principle, just as a person's can be.

Recognition of these issues by the profession considered as a social institution (as opposed to isolated individuals within it) will be necessary if they are to be adequately faced. Insofar as the profession succeeds in reassuring the public that mechanism (of sufficient complexity) is in principle capable of generating the distinctively human characteristics of subjectivity, purpose, and choice, the increasing social use of intelligent machines will present less of a threat to humane conceptions of self and society.

PLATES

Computers from the sixteenth century to the present

1. Wilhelm Schickard's Mechanical Calculator, built in 1592 in Germany. The machine could perform addition and subtraction by mechanical means, using rods, gears, and an automatic carrying mechanism. (Photograph by courtesy of IBM (UK) Ltd.)

2. The Difference Engine, constructed in 1820 by Charles Babbage, Professor of Mathematics at Cambridge University. The machine could handle quadratic functions and was to be followed by the Analytical Engine, which was operated by Jacquard loom cards. The Difference Engine would automatically produce moulds from which printing type could be cast. Support for this machine, which was far ahead of its time, was not forthcoming and it was never built.

3. Pilot model of the Ace Machine, built in 1950 at the National Physical Laboratory. Ace was probably the first machine to be used to provide an actual computing service. (Photograph Crown Copyright, National Physical Laboratory.)

4. Control console of the Leo One machine, built by J. Lyons & Co. in the early 1950s to handle accounting work. This was one of the first machines to provide an in-house data processing service. (Photograph by courtesy of ICL Ltd.)

5. The IBM 701, the first of a series available in the early 1950s which was to establish IBM as the major supplier of large computers. (Photograph by courtesy of IBM (UK) Ltd.)

6. The TX-0 computer, built at the Lincoln Laboratory, MIT, in the late 1950s. TX-0 was the first stored program computer to be made using transistors. (Photograph by courtesy of Digital Equipment Ltd.)

7. A currently available business microcomputer – the IBM PC AT. Icons – pictorial representations of functions which the machine can perform – are displayed on the screen. (Photograph by courtesy of IBM (UK) Ltd.)

8. A British competitor for the IBM AT. The XEN-i is manufactured by Apricot Computers Ltd and is claimed to be superior in performance to the AT, although compatible with it and available at a lower price. (Photograph by courtesy of Apricot Computers Ltd.)

9. The Thorn-EMI Liberator portable text processing machine, which weighs only 1.81 kg. Thorn-EMI developed this machine to follow the recommendations produced by the Civil Service for a small lightweight portable. (Photograph by courtesy of Thorn-EMI Ltd.)

10. The Cray-2 Supercomputer. The machine contains 4 central processing units and is clocked at intervals of 4.2 nanoseconds (one nanosecond equals one thousand millionth of a second). At this speed it is essential to minimise the distance between components – hence the unusual construction. The machine is only 135 cm in diameter, which presents problems for getting rid of generated heat. It is conducted away by a cascade of fluorocarbon liquid

coolant which flows through its interior. (Photograph by courtesy of Cray Computers.)

The heart of the computer

11. A silicon wafer, 10 cm in diameter, containing a number of Cellular Array Processor chips, together with four test chips. (Photograph by courtesy of ITT Ltd.)

12. Enlarged photograph of one of the Cellular Array Processor chips. This chip, which is 1.65 cm long, contains 120,000 transistors and is "fault-tolerant" – that is it will still function even if some of the transistors become faulty. The chip comprises 256 16-bit processing units. (Photograph by courtesy of ITT Ltd.)

13. The Inmos 32-bit IMS T414 transputer. This chip contains the equivalent of 200,000 transistors on a 9 mm square. A number of links, each working at 10 megabits per second (Mbps), enable it to communicate with other transputers interconnected to form an array. One hundred and fifty transputers have been incorporated in a machine called the Computing Service, which runs at a speed of 1,200 million instructions per second (mips), approaching supercomputer speeds. (Photograph by courtesy of Inmos Ltd.)

14. The Tandon disk drive with control circuits mounted on a plug-in board for a microcomputer. The drive carries a Winchester disk with a capacity of 21 megabytes. (Photograph by courtesy of Tandon Ltd.)

Computer manufacturing processes

15. An Ion implantation machine in use in the clean room at Inmos. Ion implantation changes the characteristics of the surface of semiconducting devices. (Photograph by courtesy of Inmos Ltd.)

16. The ICL Content Addressable File Store under construction. The store contains 9 printed circuit boards and is used in conjunction with the 2900 processor for extremely fast information retrieval. (Photograph by courtesy of ICL Ltd.)

Information technology in action

17. "Teleshopping" on Prestel. Items displayed on the screen may be ordered, and then paid for by credit card. (Photograph by courtesy of British Telecom plc.)

18. Bob Campbell, student journalist at Napier College, engaged in electronic publishing. He is helping to produce the publication *Village View*. (Photograph by courtesy of Apple Computers Ltd.)

19. The Kurzweil Reading Machine in action. The photograph shows a blind person at the controls listening to its output and transcribing onto an Apple II microcomputer. (Photograph by courtesy of John Bradburn, UK agent for Kurzweil Computer Products Ltd.)

20. The ICL One Per Desk voice/data microcomputer in use in an office. The machine embodies communication facilities to control two telephone lines for data or voice communication. (Photograph by courtesy of ICL Ltd.)

21. An educational system – the Sony Video Responder. The student's communication unit is shown in the foreground, printer in the centre, and display in the background. (Photograph by courtesy of Sony (UK) Ltd.)

22. At work with the Rank Cintel Retriever. Pictures or text, stored on microfilm, may be called up and displayed on the screen. Any one out of one million A4 size stored images may be retrieved and displayed in about 6 seconds. The display has a far higher resolution than a television screen so that fine detail may be examined. (Photograph by courtesy of Rank Cintel Ltd.)

23. An artist using a Quantel Paintbox – the electronic equivalent of a palette, with an almost unlimited range of colours, paper, and paintbrushes. The keyboard is used to select any colour, which can then be applied with a chosen brush thickness to paint images which appear on the screen. Different letter fonts, or stored images, may be merged with the image being drawn. (Photograph by courtesy of Quantel Ltd.)

24. Satellite Earth Station dishes at Mercury's Communications Centre in Oxfordshire. The dishes handle communications to the USA, Hong Kong and the Far East, and the Middle East and Caribbean. (Photograph by courtesy of Mercury Communications Ltd.)

25. Engineers installing a fibre optic communications cable within sight of the Tower of London. Excavations are unnecessary since Mercury has purchased the pipe network owned by the old London Hydraulic Power Co. The pipes were used to supply water under pressure for operating lifts and other machinery over a wide area. (Photograph by courtesy of Mercury Communications Ltd.)

1

2

3

4

5

6

7

8

9

10

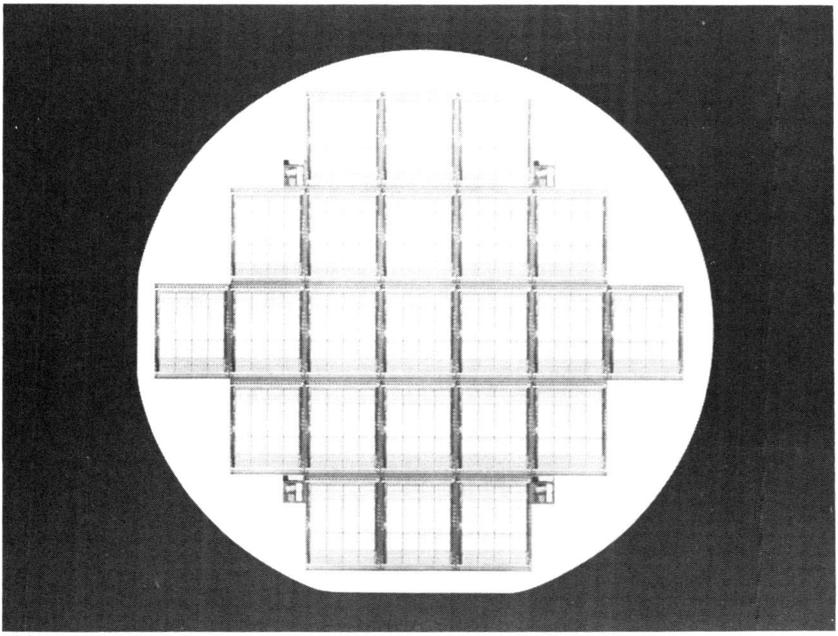

11

12

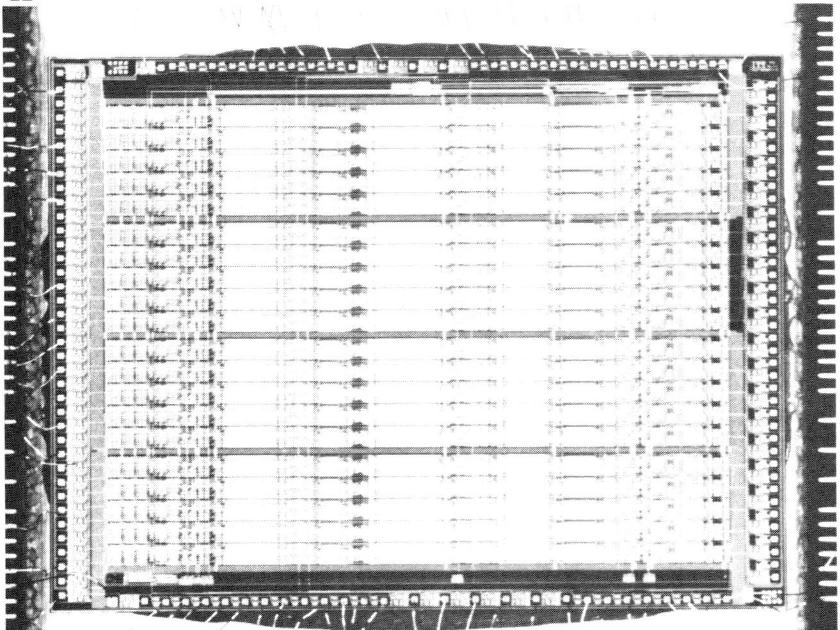

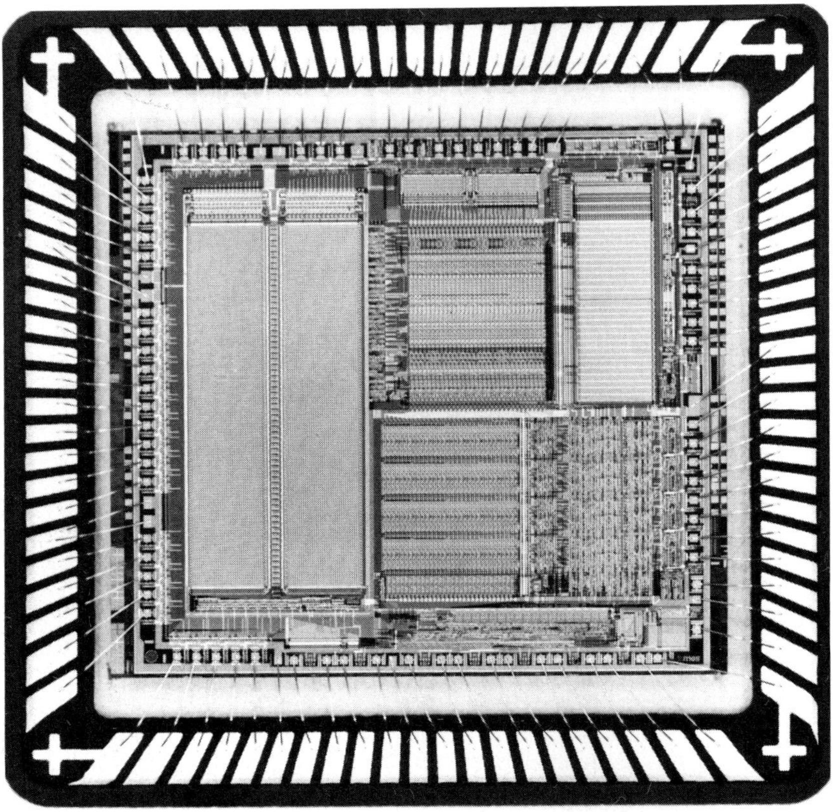

13

14

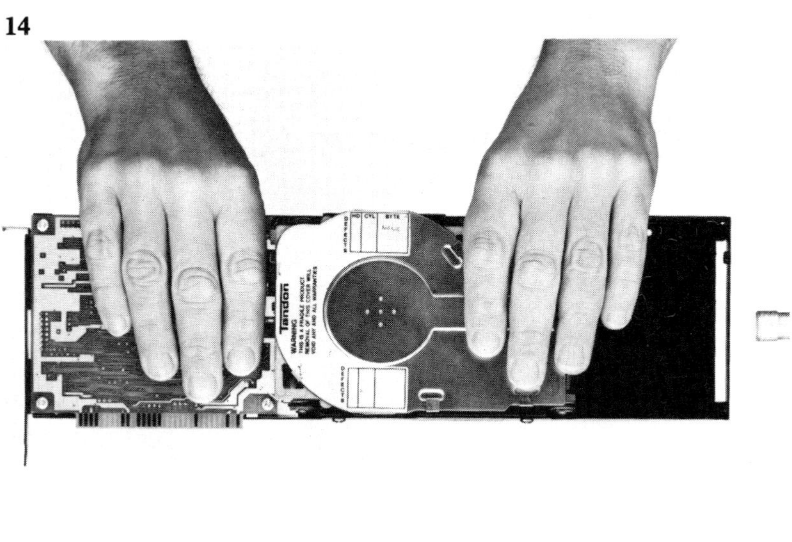
Tandon
WARNING
THIS IS A FRAGILE PRODUCT
REMOVAL OF THIS COVER WILL
VOID ANY AND ALL WARRANTIES
DEFECTS
HD
CYL

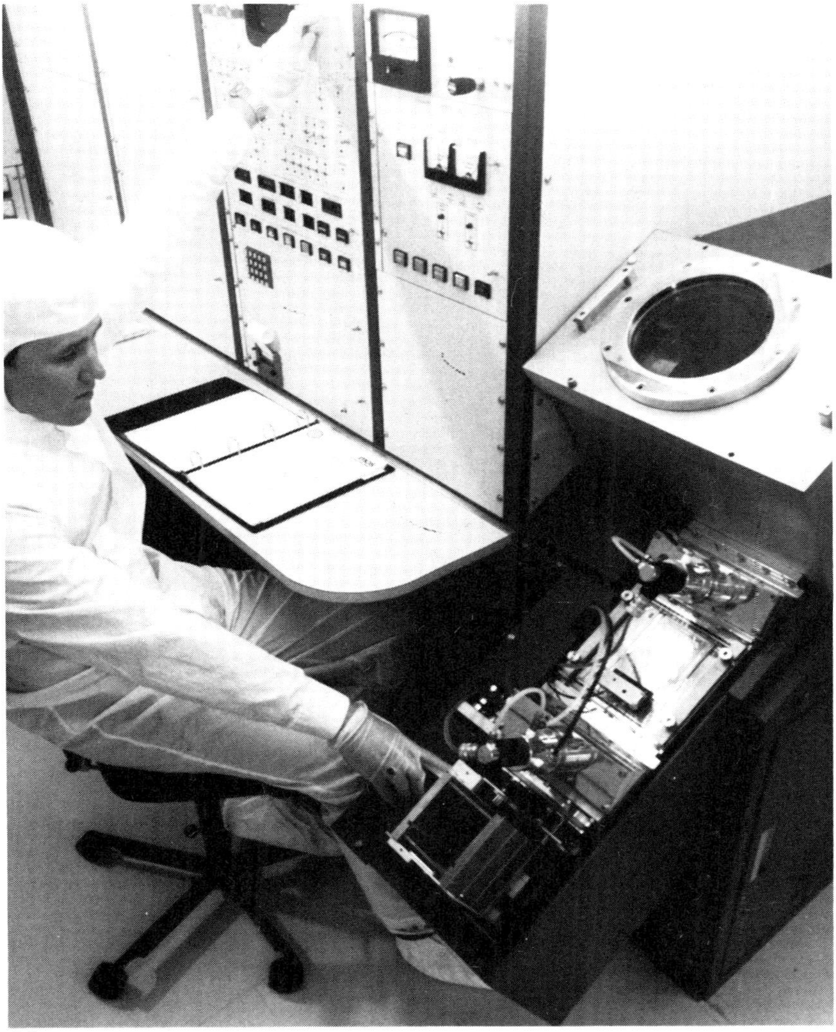

15

17

18

19

SENIORS

21

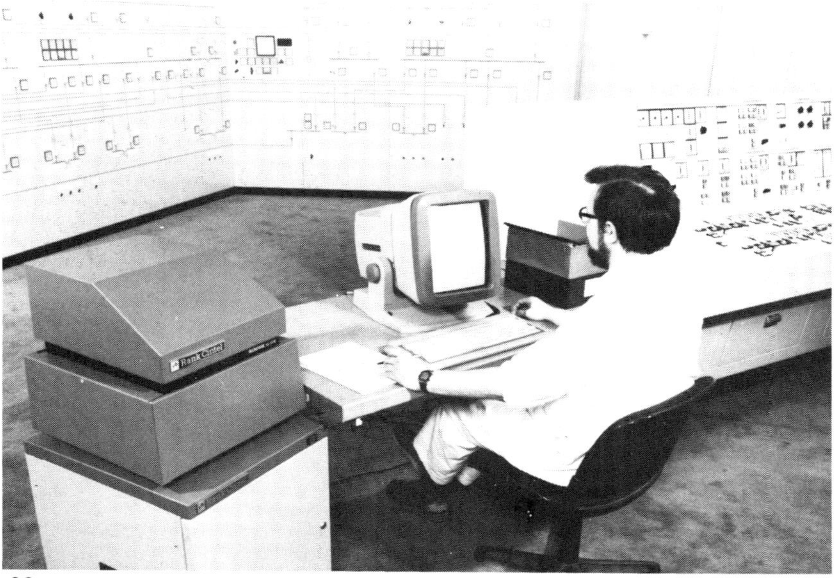

22

23

24

25